Operation Lusty:
the race for Hitler's secret technology

Graham M Simons

First published in Great Britain in 2016 by
Pen and Sword Aviation

An imprint of
Pen & Sword Books Ltd
47 Church Street
Barnsley
South Yorkshire
S70 2AS

Copyright © Graham M. Simons, 2016

ISBN 9781473847378

The right of Graham M. Simons to be identified as Author of this work has been asserted by him in accordance with the Copyright, Designs and Patents Act 1988.

A CIP catalogue record for this book is available from the British Library

All rights reserved. No part of this book may be reproduced or transmitted in any form or by any means, electronic or mechanical including photocopying, recording or by any information storage and retrieval system, without permission from the Publisher in writing.

Printed and bound in England
By CPI Group (UK) Ltd, Croydon, CR0 4YY

Pen & Sword Books Ltd incorporates the Imprints of Pen & Sword Aviation, Pen & Sword Family History, Pen & Sword Maritime, Pen & Sword Military, Pen & Sword Discovery, Wharncliffe Local History, Wharncliffe True Crime, Wharncliffe Transport, Pen & Sword Select, Pen & Sword Military Classics, Leo Cooper, The Praetorian Press, Remember When, Seaforth Publishing and Frontline Publishing

For a complete list of Pen & Sword titles please contact
PEN & SWORD BOOKS LIMITED
47 Church Street, Barnsley, South Yorkshire, S70 2AS, England
E-mail: enquiries@pen-and-sword.co.uk
Website: www.pen-and-sword.co.uk

CONTENTS

Acknowledgements 4

Introduction 5

Chapter One: Wunderwaffen 7

Chapter Two: Making use of what arrived 15

Chapter Three: The RAFwaffe 31

Chapter Four: V-for-Vengence 51

Chapter Five: The RAE Post War 73

Chapter Six: Exploiting the Enemies Developments 97

Chapter Seven: Operation Lusty and Other Missions 105

Chapter Eight: BIOS, CIOS, FIAT And Other Acronyms 201

Chapter Nine: Overcast and Paperclip 213

Chapter Ten: Across the Channel 223

Chapter Eleven: Japan 239

Chapter Twelve: Flights of Fantasy 255

Bibilography and Further Reading 259

Index 263

ACKNOWLEDGEMENTS

A project of this nature could not be undertaken without considerable help from many organizations and individuals.

Special thanks must go to Col. Richard L Upstromm and Tom Brewer from the USAF Museum, now the National Museum of the USAF, for the provision of many photographs and details.

A great amount of background material to the operations and the projects can be found in published and unpublished primary source research material in the form of memoranda, policy statements and other documents from the Army Air Corps, the USAF as provided by Lynn Gamma and all in the U.S. Air Force Historical Research Center at Maxwell Air Force Base, Montgomery, Ala. The same applies to the valuable services provided by the History Office of the Air Technical Service Command, Wright Field, Dayton, Ohio and the Public Affairs Office, Edwards Air Force Base. Much other primary source documentation is also located in the National Archives and Records Administration at College Park, Maryland.

The archives of the National Advisory Committee for Aeronautics provided access to all their relevant material, as did the archives of the Institute of Aircraft Production.

Special thanks should go to Simon Peters, Hugh Jampton, Martin Bowman and the late Peter H. T. Green for allowing me to dip into their respective photographic collections. Personal thanks must also go to David Lee, the former Deputy Director and Curator of Aircraft of the Imperial War Museum at Duxford, John Hamlin and to Vince Hemmings, the former and premier curator of the East Anglian Aviation Society's Tower Museum at Bassingbourn.

The author is indebted to many people and organisations for providing photographs for this book, many of which are in the public domain. In some cases it has not been possible to identify the original photographer and so credits are given in the appropriate places to the immediate supplier. If any of the pictures have not been correctly credited, the author apologises.

Introduction

Over the years numerous books have been written about captured German aircraft - *'War Prizes'* by Phil Butler and Kenneth S West's *'The Captive Luftwaffe'* being two excellent examples. Like many other works they look in depth at the machines - but little has been written about why the Allies studied the equipment, interrogated those who worked with it, and in many cases shipped men and machines halfway across the world to the USA.

Throughout almost the entire Second World War the Allies made use of captured Axis aircraft and equipment – either for propaganda and war-bond selling, for evaluation as to what they were flying against, or for testing to discover if and where the enemy had any advantages. From as early as 1939 the Royal Air Force and the Royal Aircraft Establishment pioneered the re-building and testing of captured enemy aircraft – flying them against their own men and machines in order to arrive at the best tactical methods of defeating the enemy. However, this could at best be only piecemeal, for all they could ever do was test what came their way.

With the entry of the United States of America into the conflict in late 1941, it fell to the British to provide the US Army Air Force with technical information as to what they were up against in the European Theatre of Operations (ETO). In the Pacific Theatre it was even harder for the Americans to evaluate the enemy because of the series of defeats they initally suffered.

In the run-up to Operation Overlord in June 1944 there were rumours of a whole range of so-called *'wunderwaffen'* - super-high-tech weapons that were far in advance of anything the Allies had which would miraculously save the Third Reich from defeat. These rumours were continually strengthened by the activities of Dr Joseph Goebbels and his Propaganda Ministry who were regularly making claims about items of equipment that were just about to be used that would force the Allies to sue for peace with the Germans.

Even to this day, it has never been revealed just how much the Allies actually believed Goebbels and his Propaganda Ministry, but nevertheless, the politicians and military men alike did take notice of all that was said, and even before the war was finally over, the Allies sent teams into both Germany and Japan hard on the heels of the front line troops in something of a mad scramble to discover and make use of the enemies' secrets. It was also clear that despite a series of political promises for sharing 'the spoils of war' - both hardware and retrieved data - with each other, there was certainly no desire or intention to actually do it!

The Allied forces had faced a whole set of frightening new German secret equipment - firstly there was the V-weapons, known in the original German as Vergeltungswaffen, meaning retaliatory or reprisal weapons. These were a particular set of long-range artillery weapons designed for strategic bombing - particularly terror bombing of cities.They comprised the V-1, a pulsejet-powered cruise missile, the V-2, a liquid-fuelled ballistic missile, and the V-3 cannon. Then there were the first jet aircraft and even rocket-powered fighters – with the imminent defeat of Germany, now the Allies could really discover what had been going on. Not only could they study the hardware, but also the plethora of ancillary equipment, items and technological achievements that went with them.

Elements within the US political and military establishment harboured resentment towards British technological advances and strong anti-British feelings that went back to the American War of Independence. Most

Americans were also at complete ideological odds with Joseph Stalin and the Soviet Union. So it should not have been surprising that when Hitler's war machine began to collapse, as far as the Americans were concerned, the race was now on to snatch as much and as many of these secrets before the other Allies – and especially the Soviet Red Army - found them.

Thus the last great battle of World War Two was fought, not for military victory but fought over and for the technology of the Third Reich. This took the form of a number of interlinked operations, including 'Operation Alsos' and its subsidiary 'Operation Big' created following the Allied invasion of Italy in September 1943 to investigate and capture or destroy what they thought could be the German nuclear energy project, 'Operation Lusty', the hunt for Nazi technologies and 'Operation Paperclip' the hunt for the German scientists.

In April 1945 American armies were on the brink of winning their greatest military victory, yet America's technological backwardness was becoming increasingly shocking when measured against that of the retreating enemy. Senior officers, including the Commanding General of the Army Air Forces, Henry Harley 'Hap' Arnold, knew all too well the seemingly overwhelming victory was, in fact, much less than it appeared. There was just too much luck involved in its outcome. Two American Army Air Force teams set out to regain America's technological edge and then exploit it for America's own ends. One went after the German technology; the other went after the Nazis' intellectual capital - their world-class scientists.

This is that story.

The thousands of reports generated in the mad dash to get as much German technology into the Allies hands as possible and used as the basis for this book resulted in a complete mis-mash of dimensions, weights, distances and speeds appearing in these documents. Feet and inches mingle with metres and centimetres. Litres mix with gallons - both Imperial and American! Distances appear in miles and kilometres, and speeds are recorded in both miles and kilometres per hour! To convert one into the other for consistency results in sets of ridiculous looking figures no matter what convention or conversion is used - therefore all dimensions that appear in this book are exactly as they appear in the primary source documentation - what was good enough for the original reports is good enough for me!

Graham M Simons
Peterborough, England
July 2015.

Chapter One

Wunderwaffen

Wunderwaffen is German for 'wonder-weapons' and was a term assigned by the Third Reich propaganda ministry to a number of revolutionary 'superweapons'.

The Reichsministerium für Volksaufklärung und Propaganda - RMVP or Propagandaministerium - known as the Reichs Ministry of Public Enlightenment and Propaganda - was a Nazi government agency to enforce Nazism ideology.

Founded upon the 1933 Machtergreifung - or seizing of power - by Adolf Hitler's National Socialist government, it was headed by Reich Minister Joseph Goebbels and was responsible for controlling the German news media, literature, visual arts, filmmaking, theatre, music, and broadcasting. As the central office of Nazi propaganda, it supervised and regulated the culture and mass media of Nazi Germany.

The English-language propaganda radio programme *Germany Calling* was broadcast to audiences in the UK on the medium wave station Reichssender Hamburg and by shortwave to the USA. The programme started on 18 September 1939 and continued until 30 April 1945, when Hamburg was finally overrun by the British Army.

Through such broadcasts, the RMVP attempted to discourage and demoralize British, Canadian, Australian and American troops and the British population within radio listening range, to suppress the effectiveness of the Allied war effort through propaganda, and to motivate the Allies to agree to peace terms leaving the Nazi regime intact and in power. Among the techniques used, the Nazi broadcasts reported on the shooting down of Allied aircraft and the sinking of Allied ships, presenting discouraging reports of high losses and casualties among Allied forces at the same time constantly broadcasting the scientific achievements of their scientists and technicians to create a soon-to-be-launched threat of the new German wonder weapons.

With the benefit of hindsight most of these weapons remained just feasible prototypes, or reached the combat theatre too late and in too insignificant numbers to have any great military effect. At the time however, those in power on the Allied side did not know that – nor could they take the risk of letting such super-weaponry fall into other hands.

This was particularly true of the USA, who not only did not trust their allies in the slightest, many of the political elite were harbouring ambitions of being the world's only super-power at the cost of the British Empire!

With the passing of time it became increasingly difficult to decipher what was fact, what was fiction and what was pure fantasy in the context of the wunderwaffen. Instead of clarification coming from the removal of the Iron Curtain and declassification of secret documents, all that has happened is that the water has become increasingly muddied by conspiracy theorists and Unidentified Flying Object fantasists.

Paul Joseph Goebbels (b. 29 October 1897 – d. May 1945) was a German politician and Reich Minister of Propaganda in National Socialist Germany from 1933 to 1945.

That said, the list of known or alleged *wunderwaffen* was huge, and could be split into numerous sections, but please remember, this is just a partial list!

Naval vessels

Graf Zeppelin – a 33,550 ton aircraft carrier that was the lead ship in a class of four carriers ordered by the Kriegsmarine, planned in the mid-1930s by Grand Admiral Erich Raeder as part of the Plan Z rearmament programme after Germany and Great Britain signed the Anglo-German Naval Agreement. The carrier would have had a complement of forty two fighters and dive bombers.

A combination of political infighting between the Kriegsmarine and the Luftwaffe, disputes within the ranks of the Kriegsmarine itself and Adolf Hitler's waning interest all conspired against the carriers. A shortage of workers and materials slowed construction still further and, in 1939, Raeder reduced the number of ships from four to two. Even so, the Luftwaffe trained its first unit of pilots for carrier service and readied it for flight operations. With the advent of the war, priorities shifted to U-boat construction; one carrier, *Flugzeugträger B*, said by some sources to have been named *Peter Strasser*, - the chief commander of German Imperial Navy Zeppelins during World War I - was broken up on the slipway while work on the other, *Flugzeugträger A* (christened *Graf Zeppelin*) was continued tentatively but suspended in 1940. The air unit scheduled for her was disbanded at that time.

German Aircraft Carrier I– This was a planned conversion of the transport ship *Europa* during World War Two. The loss of the battleship *Bismarck* and near torpedoing of her sistership *Tirpitz* in May 1941 and March 1942, respectively spurred the Kriegsmarine to acquire aircraft carriers. *Europa* was one of several vessels selected for conversion into auxiliary aircraft carriers. As designed, the ship would have had an air complement of 24 Bf.109T fighters and 18 Ju.87C Stuka dive-bombers.

Somewhat ironically, the *Graf Zeppelin* was captured by the Russians at the end of the war at Swinemünde and used as a Headquarters for the Commission analysing captured German equipment.

H-class battleship – a series of proposals for battleships surpassing both the US Navy's Montana-class battleships and the Imperial Japanese Navy's Yamato-class battleships in armament, culminating in the H-44, a 140,000 ton battleship with eight 20 inch guns. Two only laid down; both were scrapped on slipways.

U-boats

Rocket U-boat – a planned ballistic missile submarine; project abandoned.
Type XVIII U-boat – a U-boat designed to use air-independent propulsion; several were under construction when the war ended
Type XXI U-boat 'Elektroboot' (Electric boat) – the first U-boat designed to operate completely submerged. 118 were built

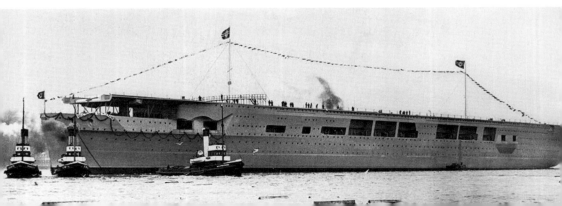

The German aircraft carrier *Graf Zeppelin* after its launch in December 1938.

Type XXIV U-boat – a planned U-boat designed to use air-independent propulsion
Type XXVI U-boat – designed to use air-independent propulsion.
Type XXII U-boat – designed to use air-independent propulsion.
Type XXIII U-boat ('Elektroboot') – a U-boat designed for coastal missions; 67 were built
Type XXV U-boat – a planned all-electric U-Boat designed for coastal missions
Type XI U-boat – designed to carry the Arado Ar 231 collapsible floatplane; four were laid down but cancelled at the outbreak of World War Two

Armoured vehicles

Flakpanzer 'Kugelblitz' (Ball Lightning) – a self-propelled anti-aircraft gun.
Sturer Emil – an experimental tank destroyer.
Landkreuzer P. 1000 'Ratte' (Rat) – a planned super-heavy tank, weighing 1000 metric tons and armed with two 280mm cannons, 128mm anti-tank gun, eight 20mm flak guns and two 15mm heavy machine guns.
Landkreuzer P. 1500 'Monster' – a proposed super-heavy self-propelled gun, weighing 1500 metric tons and armed with the 800mm Schwerer Gustav/Dora gun
Panzer VII 'Löwe' (Lion) – a planned super-heavy tank, weighing 90 metric tons and armed with a 105mm cannon.
Panzer VIII 'Maus' (Mouse) – a super-heavy tank, weighing 180 metric tons and armed with two cannons of 128mm and 75mm calibre, two operable prototypes completed.
Panzerkampfwagen E-100 – a planned super-heavy tank, weighing 140 metric tons and armed with either 128, 149 or 170mm cannon.

Aircraft

Junkers Ju.322 'Mammut' (Mammoth) – a flying wing heavy transport glider.
Focke-Achgelis Fa.269 – a planned tilt-rotor VTOL fighter.
Focke-Wulf Ta.152 – a high-altitude interceptor.
Focke-Wulf Ta.400 – a planned Amerika Bomber candidate with six radial engines and two jet engines with a range of 13,000 km in bomber configuration
Heinkel He.111Z – a five engined Zwilling (twin fuselage) aircraft created by combining two He.111s and designed to tow large gliders
Heinkel He.274 – a high altitude heavy bomber with four in-line engines with a range of 3,440 km, two completed by France after the war
Heinkel He.277 – a planned, advanced long range bomber design, never built as a

Designed for use on U-boat 'cruisers', like the Type XI B, the Arado Ar 231 was a light parasol-wing aircraft. It was powered by a 160 hp Hirth HM 501 inline engine, weighed around 1,000 kg, and had a 10 metre wingspan. The design led to a simple and compact aircraft that could be fitted into a storage cylinder only two metres in diameter. One unusual feature was an offset wing design, with the right wing root attaching to the wing's tilted centre section and lower than the left wing root, to allow the wings to be quickly folded up.

Panzerkampfwagen VIII Maus ('Mouse') was a German super-heavy tank completed in late 1944. It is the heaviest fully enclosed armoured fighting vehicle ever built. Only two hulls and one turret were completed before the testing grounds were captured by the advancing Soviet forces.

complete aircraft, evolved to be an Amerika Bomber candidate, to be powered with four BMW 801 radial engines and up to 11,000 km range.
Junkers Ju.390 – an Amerika Bomber candidate with six radial engines with a range of 9,700 km. Two airworthy prototypes built and flown.
Junkers Ju.488 – a heavy bomber with four radial engines with a range of 3,395 km.
Messerschmitt Me.264 – an Amerika Bomber candidate with four inline or radial engines and a range of 15,000 km, three airworthy prototypes built and flown.
Messerschmitt Me.323 'Gigant' (Giant) – a heavy transport with six engines.
Arado Ar.234 – the first operational turbojet bomber and reconnaissance aircraft.
Arado E.555 – a planned jet-powered Amerika bomber.
Arado E.560 – a series of tactical bomber projects.
Bachem Ba.349 'Natter' (Adder) – a rocket-powered vertical takeoff interceptor.
Blohm & Voss P.178 – a turbojet dive bomber.
DFS.194 – a rocket-powered experimental aircraft.
DFS.228 – a rocket-powered high altitude reconnaissance aircraft.
DFS.346 – a rocket-powered research aircraft.

A model of the Messerschmitt P.1106 swept wing jet fighter

The Fa.269 was a tiltrotor VTOL single seat fighter. The machine did not make it into production since the war ended before it could be produced. It went through extensive wind tunnel testing and a full scale prototype was supposedly built, but was destroyed by bombing, so it did not survive the war.

Three views of the little-known Lippisch P.13b. This appears to be one of the most controversial aircraft of World War Two - not due to any special paper project design discovered apart from the others, but due to the claims that the United States deliberately kept classified documents that may prove that the aircraft was actually built in 1945 and was the first to break the sound barrier.

Fieseler Fi.103R 'Reichenberg' – a manned version of the V-1 flying bomb.
Focke-Wulf 'Triebflügel' (Powered Wings) – a tip jet rotorcraft, tailsitter interceptor.
Focke-Wulf Ta.183 'Huckebein' – a planned swept wing turbojet fighter.
Focke-Wulf Ta.283 – a planned swept wing ramjet and rocket-powered fighter.
Heinkel He.162 'Volksjäger' (People's Fighter) – a turbojet fighter.
Heinkel He.176 – the world's first liquid-fuelled rocket-powered experimental aircraft
Heinkel He.178 – the world's first experimental turbojet aircraft to fly.
Heinkel He.280 – the first turbojet fighter design, prototypes only.
Heinkel He.343 – a four engined jet bomber based on the Arado Ar 234.
Henschel Hs.132 – a planned turbojet dive bomber and interceptor.
Horten Ho.229 – a turbojet flying wing jet fighter/bomber.
Horten H.XVIII – a planned flying wing jet bomber based on the Horten Ho 229.
Junkers EF.128 – a planned turbojet fighter.
Junkers EF.132 – a planned turbojet bomber.
Junkers Ju.287 – a forward-swept wing turbojet bomber.
Lippisch P.13a – a planned supersonic ramjet delta wing interceptor.
Lippisch P.13b – a ramjet delta wing interceptor developed from the Lippisch P.13a.
Messerschmitt Me.109TL – a turbojet fighter designed as an alternative to the Me.262.
Messerschmitt Me.163 'Komet' (Comet) – a rocket-powered fighter.
Messerschmitt Me.262 'Schwalbe' (Swallow) – a turbojet fighter/bomber.
Messerschmitt Me.263 – a rocket-powered fighter developed from the Me 163.
Messerschmitt P.1101 – a variable-sweep wing turbojet fighter.
Messerschmitt P.1106 – a jet fighter based on the Messerschmitt Me P.1101.
Skoda-Kauba Sk P.14 - a ramjet-powered emergency fighter.

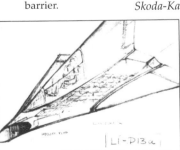

In late 1943 Deutsche Arbeitsfront Director, Otto Lafferenz, proposed a towable watertight container which could hold an A4 rocket. The design was a container of 500 tons displacement to be towed behind a U-boat. Un-manned and unpowered, it was towed to within range of its target, flooded into an upright position, and the missle then launched. Three of these vessels were ordered in late 1944, but only one was built. The project was dubbed Projekt Schwimmweste and the containers themselves referred to by the codename Prüfstand XII. Work on the containers was carried out by the Vulkanwerft, and a single example was completed by the end of the war, but never tested with a rocket launch. The design looks remarkably like a maritime version of the underground missile silos used both by the Americans and the Soviets during the Cold War.

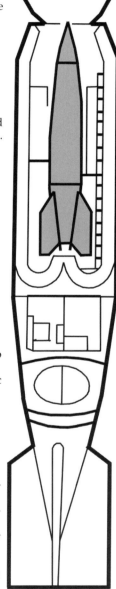

Sombold So.344 - a rocket-powered plane with a detachable explosive nose.
Silbervogel (Silverbird) – planned sub-orbital antipodal bomber.
Zeppelin Fliegende Panzerfaust – a rocket-powered very-short-range interceptor.
Zeppelin Rammer – a rocket-powered ramming interceptor.

Helicopters
Flettner Fl.184 – a night reconnaissance and anti-submarine autogyro.
Flettner Fl.185 – an experimental helicopter.
Flettner Fl.265 – an experimental helicopter.
Flettner Fl.282 'Kolibri' (Hummingbird) – a reconnaissance helicopter
Focke-Achgelis Fa.223 'Drache' (Dragon) – an anti-submarine, search and rescue, reconnaissance, and freight helicopter, based on the pre-war Fw 61.
Focke-Wulf Fw.61 – an experimental helicopter.

Bombs, rockets and explosives
German nuclear energy project
Schwerer Gustav (Heavy Gustav) – an 800mm railway gun
V-3 cannon 'Hochdruckpumpe' – 'High Pressure Pump'; a supergun
A1 – the first German experimental rocket
A2 – an experimental rocket, gyroscopically stabilised
A3 – an experimental rocket with an inertial guidance system.
A4/V-2 – a ballistic missile.
A4-SLBM – a planned submarine-launched ballistic missile.
A4b - Longer range version of the A4 rocket.
A5 – an experimental reusable rocket.
A6 – an improved A4b rocket.
A7 – an improved A4 rocket.
A8 – a planned submarine-launched ballistic missile.
A9 Amerikarakete – an intermediate-range ballistic missile to be used to strike the eastern United States.
A10 – a lower stage for the A9 to upgrade it to an intercontinental ballistic missile.
A11 – a planned satellite launcher.
A12 – a rocket, capable of putting 10 metric tons into low Earth orbit.
Enzian – a surface-to-air missile with infrared guidance.
Feuerlilie F-25 'Fire Lilly' – a surface-to-air missile.
Feuerlilie F-55 'Fire Lilly' – a two-stage, supersonic surface-to-air missile.
V-1 flying bomb/Fieseler Fi 103/Vergeltungswaffe 1 – the first cruise missile.
Fliegerfaust 'Pilot Fist' or 'Plane Fist' / *Luftfaust* 'Air Fist' – the first man-portable air-defence system (MANPADS).
Fritz X – an unpowered air-launched, manual command to line of sight-guided anti-ship missile.
Henschel Hs.117 Schmetterling 'Butterfly' – a manually guided surface-to-air missile.
Henschel Hs.117H – a manually guided air-to-air missile.
Henschel Hs.293 – a manual command to line of sight guided air-to-ship missile.

Henschel Hs.294 – a manual command to line of sight guided air-to-ship missile/torpedo.
Henschel Hs.298 – an air-to-air missile.
R4M Orkan 'Hurricane' – an unguided air-to-air rocket.
Rheinbote 'Rhine Messenger"- the first short-range ballistic missile.
Rheintochter 'Rhinedaughter' – a manually guided surface-to-air missile.
Ruhrstahl X-4 – a wire-guided air-to-air missile designed for the Ta 183.
Taifun 'Typhoon' – a planned unguided surface-to-air missile.
Wasserfall Ferngelenkte Flakrakete – a supersonic surface-to-air missile.
Werfer-Granate 21 – a heavy-calibre unguided air-to-air rocket.
G7es/Zaunkönig T-5 – acoustic homing torpedo used by U-boats.

Orbital
Sun gun – a parabolic mirror in orbit designed to focus sunlight onto specific locations on the Earth's surface.

Guns
Jagdfaust – an automatic airborne anti-bomber recoilless rifle for use on the Me 163.
Mauser MG 213 – a 20 mm aircraft mounted revolver cannon.
Mauser MG 213C – a 30 mm aircraft mounted revolver cannon.
Sturmgewehr 44 – the first assault rifle.
Krummlauf – a curved barrel for the StG44.
Sturmgewehr 45 – prototype.

The Heavy Gustav railway gun.

In 1929, the German physicist Hermann Oberth developed plans for a space station from which a 100 metre-wide concave mirror could be used to reflect sunlight onto a point on the earth.

During World War Two, a group of German scientists at a research centre in Hillersleben began to expand on Oberth's idea of creating a superweapon that could utilize the sun's energy. This 'sun gun' would be part of a space station 5,100 miles above Earth. The scientists calculated that a huge reflector, made of metallic sodium and with an area of 3.5 square miles, could produce enough focused heat to make an ocean boil or burn a city. After being questioned by Allied officers, the Germans claimed that the sun gun could be completed within 50 or 100 years.

Directed-energy weapons

Rheotron/betatron - Among the directed-energy weapons the Nazis investigated was the X-ray beam developed under Heinz Schmellenmeier, Richard Gans and Fritz Houtermans. They built an electron accelerator called Rheotron (invented by Max Steenbeck at Siemens-Schuckert in the 1930s, these were later called betatrons by the Americans) to generate hard x-ray synchrotron beams for the RLM. The machine worked by interrupting the magnetos of engines in Allied bombers and brought aircraft down to lower altitudes into the reach of flak batteries.

Norwegian born Dr Rolf Wideroe wrote in his autobiography that he worked on a particle accelerator X-Ray transformer for this project at Hamburg in 1943. The Philips subsidiary Valvo also participated and much of the engineering was performed by CHF Muller & Co. Wideroe later rescued the device from the rubble of Dresden and delivered it to General Patton's 3rd Army at Burggrub on 14 April 1945.

Röntgenkanone - another approach was from Ernst Schiebold who developed this from 1943 at Großostheim. This employed a particle accelerator cupped from beneath by a Beryllium parabolic mirror with a bundle of nine beryllium rods as an anode at its core. The entire device was steerable at Allied bomber formations. The Company Richert Seifert & Co was largely responsible for its manufacture.

Given the scale and scope of projects and technology suspected of either being already in the hands of the Germans, or being actively worked on and close to fruition, it is not surprising that the Allied military machine did all they could as the war drew to its conclusion to get their hands on it all.

But investigations had started at least four years earlier.

Chapter Two

Making Use Of What Arrived

On many occasions during the Second World War German aircraft literally fell upon the United Kingdom, and they subsequently arrived at experimental establishments in various stages of wreckage or disrepair. It was by no means unusual for them to be very much the worse for wear, but sometimes good fortune provided specimens in complete working order, or suffering only minor damage.

Other Axis aircraft were also captured and evaluated. The testing and evaluation of captured aircraft and equipment in the United Kingdom was undertaken at various research establishments and specialist Service units, but came under the overall control of a department within the Air Ministry. Termed 'AI-2', this Air Ministry intelligence branch was responsible for the analysis of enemy aircraft production.

Factors in securing a good flow of enemy machines were, firstly, as shoot-downs from the fighter and anti-aircraft defences, but from the scientific viewpoint machines obtained in this manner were not over popular since they tended to have been inflicted with rather drastic damage; the second, and much better source was primarily due to bad navigation, fuel shortage, engine failure, and defection.

Naturally, the Royal Aircraft Establishment (RAE) at Farnborough came in the first category and was engaged in this function from 1939 to 1946. Farnborough was the premier centre for flight testing in the United Kingdom, with a reputation of being one of the world leaders in its field.

It's origins date back to the spring of 1906, when the Army Balloon Factory, which was part of the Army School of Ballooning relocated, from Aldershot to

The remains of a Messerschmitt Me.110 shot down by British fighters over Essex are removed by a squad of British soldiers on 3 September 1940.
(Peter H T Green Collection)

Members of the British Home Guard pose in front of this downed Junkers Ju.87 'Stuka' that bears the numbers 5167 on its rudder. The aircraft appears to be undamaged. *(Peter H T Green Collection)*

the edge of Farnborough Common in order to have enough space for experimental work. Superintendent of the Balloon Factory was Colonel John E Capper, Royal Engineers.

In the winter of 1907 American Samuel F Cody, who had been making balloons and man-carrying kites as the 'Chief Instructor of Kiting' at the Factory - came to build his 'Army Aeroplane No.1'. In September 1908 Cody made the first aeroplane flight in Britain at Farnborough when his machine made a free flight of two hundred and ten feet.

In December 1909, His Majesty's Balloon Factory and its associated 'Balloon Section, Royal Engineers' was split in two. Colonel Capper was appointed as Commander of the Balloon Section, which became the embryo of the later Royal Flying Corps (RFC), and his job as Superintendent of the Factory was taken by Mervyn O'Gorman, a consulting engineer.

By 1911, the Factory became the Army Aircraft Factory, although its main pre-occupation was initially the production of Army airships. At this time Assistant Engineers of Design, Physics and Machine Shops were appointed, laying the foundations for the later work of the Establishment. In 1912 it was further renamed as the Royal Aircraft Factory and was charged, amongst other duties, with *'…tests with British and foreign engines and aeroplanes; experimental work'.*

Among its designers was Geoffrey de Havilland, who later founded his own company, John Kenworthy who became chief engineer and designer at the

Troops inspect the burnt out remains of Heinkel He.111 of KG27 (1G+L), which came down on the south-east coast on England on 13 July 1940.

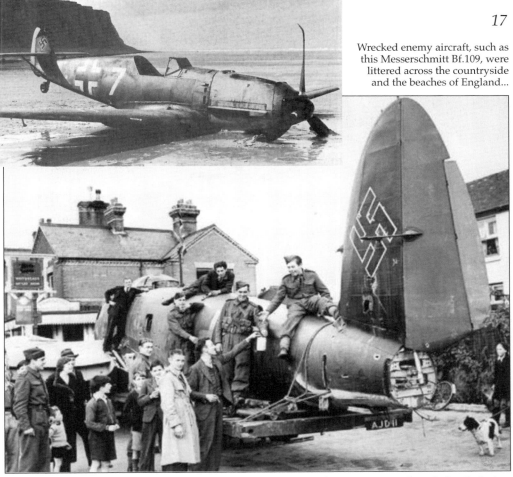

Wrecked enemy aircraft, such as this Messerschmitt Bf.109, were littered across the countryside and the beaches of England...

...where they became a centre of attention. Specialist units removed the wreckage - such as this Heinkel He.111 - for further study. *(Peter H T Green Collection)*

Austin Motor Company in 1918 and who went on to found the Redwing Aircraft Co in 1930, and Henry Folland – later chief designer at Gloster Aircraft Company, and founder of his own company, Folland Aircraft. One of the designers in the engine department was Samuel Heron, who later went on to invent the sodium-filled poppet valve, instrumental in achieving greater powers from piston engines. While at the RAF, Heron designed a radial engine that he was not able to build during his time there, but upon leaving the RAF he went to Siddeley-Deasy where the design, the RAF.8, was developed as the Jaguar. Heron later moved to the United States, where he worked on the design of the Wright Whirlwind.

Other engineers included Major F.M. Green, G.S. Wilkinson, James E. 'Jimmy' Ellor, Prof. A.H. Gibson, and A.A. Griffith. Both Ellor and Griffith would later go on to work for Rolls-Royce Limited.

During the early years of the First World War, the Royal Aircraft Factory was in almost a monopoly situation with regard to the design of aircraft for the RFC, such that it produced not only prototype aircraft designs but also supervised the production of those designs which were accepted for military service. This involved setting up production by a large number of privately-owned aircraft industry sub-contractors to the government, to the exclusion of aircraft designed by the private aircraft industry. The Factory also became involved in the production of its own aircraft and engine designs.

Activities extended not only to aircraft design and production, but also

The cockpit section of Ju.88A-1 7A+FM under examination ar the RAE. The aircraft made a wheels up landing at RAF Oakington on 19 September 1940. *(John Hamlin Collection)*

to the use of wind-tunnels and to the development of materials, instruments, propellers, and stressing methods that included the testing of aircraft to destruction.

The monopoly held by the Factory was criticised and in 1916 an official enquiry was set up which eventually ruled that the Factory should no longer be concerned with the design and construction of aircraft, but instead concentrate on the theoretical and consultative aspects.

With the creation of the Royal Air Force (RAF) on 1 April 1918, the Royal Aircraft Factory was renamed as the Royal Aircraft Establishment.

In 1924 a further official Committee was set up to report on the organisation of the Establishment, and this - the Halahan Committee - confirmed the primary function of the RAE as providing 'a full-scale aeronautical laboratory for the Air Ministry' and defined its main activities as:
- development work on experimental aeroplanes and engines .
- testing of experimental instruments and accessories .
- development of special flying instruments for which there is little commercial demand .
- investigations of failures; liaison with contractors' research.
- technical supervision of the construction of experimental machines.
- stressing of new types of machine, approval of designs and the issue of airworthiness certificates.
- the issue of certain technical publications.

In the inter-war period, subject to the severe financial restraints of the time, much work was done in the field of aero engine development, especially

This Junkers Ju.88 crash-landed during the Battle of Britain and is seen here guarded by members of the British Army. From the condition of the propellers the machines must have slid backwards during its final landing. *(Peter H T Green Collection)*

A police constable and Army private investigate this downed Heinkel He.111. *(Peter H T Green Collection)*

Local newspapers carried reports of enemy aircraft being put on display. This Heinkel 111, coded 4H+FM and bearing signs of battle-damage, was escorted by police through Peterborough. It was common to sheet over the cockpit area, especially if it showed signs that the crew had been killed. *(author's collection)*

with regard to engine superchargers and automatic engine controls. Other areas included aircraft fire extinguishing research, a comprehensive programme of work on aerofoils and wind-tunnel testing of models for general aerodynamic research, catapult launching and pilotless aircraft developments, and pioneering work on Gas Turbine development for aircraft applications. The whole of the RAF radio communications network was reconstructed under RAE supervision. These tasks were undertaken with a technical staff of about 150 people, until the numbers began to build up from 1934 with the prospect of a European war looming.

With the outbreak of the Second World War, the technical staff of the RAE became increasingly involved in the investigation of crashed enemy aircraft and captured equipment. The investigation work drew on the experience of all departments of the Establishment. This experience had been built up during the formative years of peace.

These activities resulted in the issue of the 'Enemy Aircraft' (EA) series of special technical reports. From June 1940, when the Experimental Flying Department became involved in flight tests of a captured Messerschmitt Bf.109E fighter, these reports included flight test results, engineering appraisals and descriptions of complete aircraft received at the Establishment or examined 'on site' at other experimental units or at the locations where they had been shot down. The main appraisal work was co-ordinated by Mr W. Sutcliffe, who was put in charge of an Enemy Aircraft and Engine Section of the Mechanical Test

Two views of this badly damaged Junkers Ju.88 from 7th Staffel, KG30 that was put on display in the grounds of Kings School, Park Road Peterborough. It had been shot down near Driffield in Yorkshire on 15 August and was displayed for 'War Week' in late 1940. Local legend has it that members of the public, including King's School pupils, could climb into the aircraft for a closer look, once they had made their 15 shilling purchase!
[both author's Collection]

Department of the Establishment. Staff from this Section were available at short notice to carry out on-site examinations of enemy aircraft which had crashed or been shot down in the United Kingdom.

The majority of complete German and Italian aircraft captured after force-landing (or landing in error) in England were brought to Farnborough for examination and test. Later in the war examples of aircraft types or variants which had already been assessed at the RAE were sometimes sent to 1426 Flight or other units without being moved to Farnborough, although even in these cases technical specialists from the RAE often examined them and wrote reports describing any new features or equipment found.

The main flight test work was carried out by the Aerodynamics Flight of the Experimental Flying Department, although the Wireless & Electrical Flight also flew many hours, especially with investigations of radar-equipped aircraft later in the war.

With the coming of World War Two, the first few enemy aircraft shot down over Britain were objects of the greatest interest and a team of experts was detailed for the sole task of examining the wrecks. The supply soon became more plentiful and it was possible, in some instances, to repair different types

One of a number of official air-to-air photographs of He.111 H-1 AW177 - formerly 1H+EN of IIKG26. The Lion Geschwader emblem is seen below the pilot's side window. *(Peter H T Green Collection)*

and put them through flight tests to determine their capabilities. After the Battle of Britain and the subsequent bombing raids, the RAE in fact possessed a not insignificant set of Luftwaffe machines, and it was a common enough event to see mock dog-fights between British and German aircraft over Farnborough.

The responsibility for flying captured aircraft rested on the shoulders of such distinguished test pilots as Group Captain H. J. Wilson, among others.

Known as 'Willie' Wilson (b. 28 May 1908 – d. 5 September 1990) Hugh John Wilson was a senior Royal Air Force officer. He served as the RAF's main chief test pilot for captured enemy aircraft.

On 13 September 1929, Wilson joined the RAF within the General Duties Branch on a short service commission. He trained at 5 Flying Training School and was later posted to 111 Squadron. Exactly five years after joining the RAF in 1934, he was placed on the Reserve of Air Force Officers. Whilst on this reserve list he passed a conversion course on flying boats and also as a flying instructor.

In the mid-1930s he worked as a test pilot for Blackburn Aircraft and was, in 1938, the first to fly the Blackburn Roc. He later worked at the Royal Aircraft Establishment at Farnborough as a civil test pilot.

Recalled to active service with the RAF in 1939, Wilson was the Commanding Officer of the Aerodynamic Flight of the RAE at RAF Farnborough.

An Italian Arrival.

A little known fact about the Battle of Britain was that the UK was attacked by the Italian Regia Aeronautica. This occurred when a group of FIAT CR.42 biplane fighters - suggested by some records as being around sixty - from 95[a] Squadriglia Caccia Terrestre of the 18° Gruppo, 56° Stormo attached to the Corpo Aereo Italiano (Italian Air Corps) under the command of the Luftwaffe's Luftflotte 2 escorted between fifteen and twenty Caproni bombers who attacked the south east of England. This unit, based at Maldegen in Belgium, had at least one machine – MM5701 coded 13-95 - shot down by RAF Hurricanes from Martlesham Heath on 11 November 1940. The aircraft made a force-landing

Ju.88 G-6 'AM 3' was delivered by road to 6 MU on 23rd July 1945 for preparation for display in Hyde Park London. It returned to the MU until 11 December, when it was returned to RAE Farnborough. Its subsequent fate is unknown. *(Peter H T Green Collection)*

near Orfordness with a fractured oil-pipe but was otherwise undamaged. Contemporary reports suggest that the Hurricanes shot down eight bombers and five fighters.

The CR.42 was taken by road to Martlesham Heath and then, on 27 November, to Farnborough. The fighter was subsequently allotted the RAF serial BT474. Only limited flights were made at RAE by Squadron Leader L. D. Wilson before it was flown to the AFDU at Duxford by Wing Commander I. R. Campbell-Orde on 28 April 1941, to enable that unit to develop tactics against the type, which was the standard Italian fighter in use in the Middle East theatre, by mock combat with various types of fighter in service with the RAF.

During 1943 the AFDU lost interest in the type because of its obsolescence and the aircraft was selected for storage as potential museum material.

Between 1941 and the end of the war Wilson was the RAF's main test pilot on all captured enemy aircraft. Flying these aircraft from RAF Farnborough (they had been repainted with RAF roundels), Wilson would evaluate their handling and performance.

One such 'acquisition' was the capture of a Focke-Wulf Fw.190A-3 of III/JG 2. The story behind its arrival is most intriguing. Luftwaffe pilot Oberleutnant Armin Faber mistook the Bristol Channel for the English

Two of the downed Italian aircraft: a twin-tailed Caproni bomber and a biplane FIAT CR.42 fighter.

FIAT CR.42 MM5701 95ª Squadriglia Caccia Terrestre of the 18° Gruppo, 56° Stormo attached to the Corpo Aereo Italiano seen on a test flight in RAF markings as BT474.

Heinkel He.115B-1 BV186. This machine had a somewhat varied career, having been captured from the Luftwaffe by the Norwegians during the fighting for Norway and taken over by the *Marinens Flyvevaesens*. It was operated by the Royal Norwegien Navy as '64' until evacuated to the UK in June 1940, being flown then by British Overseas Airways Corporation. It was then flown to Malta on October 1941, being destroyed at Kalifrana by strafing early in 1942. *(John Hamlin Collection)*

Channel and landed his Focke-Wulf 190 intact at RAF Pembrey in south Wales, thinking he was in France! This machine was the first Fw-190 to be captured by the Allies and was tested to reveal any weaknesses that could be exploited.

It all began when in June 1942 Oberstleutnant Armin Faber was Gruppen-Adjutant to the commander of the III fighter Gruppe of *Jagdgeschwader 2* (JG 2) based in Morlaix in Brittany. On 23 June, he was given special permission to fly a combat mission with 7th Staffel.

The Fw.190 had only recently arrived with front line units and its superior performance had caused the Allies so many problems that they were considering mounting a commando raid on a French airfield to capture one for evaluation.

7 Staffel was scrambled to intercept a force of twelve Bostons on their way back from a bombing mission; the Bostons were escorted by three Czech-manned RAF squadrons, 310, 312 and 313. A fight developed over the English Channel with the escorting Spitfires, during which Faber was attacked by Czechosolvakian Sergeant František Trejtnar of 310 Squadron. In his efforts to shake off the Spitfire, Faber flew north over Exeter in Devon.

Messerschmitt Bf.109E-3 landed in error due to fog in an orchard at Woerth, in Bass-Rhin in France. It was transferred to the CEV (or *Centre d'Essais en Vol* - flight test centre) at Bricy and painted in French colours, but retained its 1/JG76 insignia, construction number and 'White 1'. The aircraft was handed over to the RAF at Amiens on 2 May 1940 and was flown to A&AEE Boscombe Down via Chartres and Tangmere the next day and allocated the RAF serial AE479.

The cockpit of the aircraft. Translations have been painted on some of the instruments. *[both Hugh Jampton Collection]*

After much high-speed manoeuvring, Faber, with only one cannon working, pulled an Immelmann turn into the sun and shot down his pursuer in a head-on attack.

Trejtnar bailed out safely, although he had a shrapnel wound in his arm and sustained a broken leg on landing; his Spitfire crashed near the village of Black Dog, Devon. Meanwhile, the disorientated Faber now mistook the Bristol Channel for the English Channel and flew north instead of south. He turned towards the nearest airfield - RAF Pembrey. Observers on the ground could not believe their eyes as Faber waggled his wings in a victory celebration, lowered the Focke-Wulf's undercarriage and landed.

The Pembrey Duty Pilot, Sergeant Jeffreys, immediately grabbed a signal

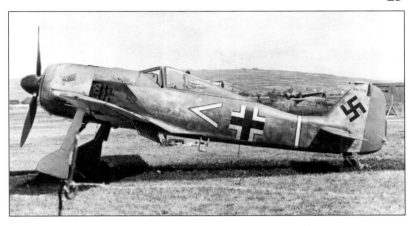

Faber's Focke-Wulf Fw.190A-3 of III/JG 2 at RAF Pembrey, June 1942.

Faber's captured Focke Wulf Fw.190A-3 at the Royal Aircraft Establishment, Farnborough, with the RAE's chief test pilot, Wing Commander Hugh J 'Willie' Wilson at the controls, August 1942. The machine would later have a yellow 'P' in a circle painted behind the roundel denoting 'prototype'.

pistol and ran from the control tower and jumped onto the wing of Faber's aircraft as it taxied in. Faber was apprehended and later taken to RAF Fairwood Common by Group Captain David Atcherley for interrogation.

Faber's aircraft was a Fw.190A-3 with the Werknummer 313. It was the only fighter configuration to be captured intact by the Allies during the war. All other captured aircraft were either of the long range bomber or fighter bomber configuration.

Group Captain Hugh Wilson, the pilot mainly responsible for test flying captured enemy aircraft, was asked to fly 313 from RAF Pembrey to RAF Farnborough under the guarantee that he would not crash. This was an impossible guarantee to give, so the aircraft was dismantled and transported via lorry instead.

At Farnborough, the Fw-190 was repainted in RAF colours and given the RAF serial number MP499 and a 'P' for prototype. Testing and evaluation commenced on 3 July 1942 at the RAE. Roughly nine flying hours were recorded, providing the Allies with extremely valuable intelligence.

After ten days it was transferred to the Air Fighting Development Unit at RAF Duxford for further tactical assessment, where it was flown in mock combat trials against the new Spitfire Mk.IX, providing the RAF with methods to best combat the Fw.190A with their new fighter.

The Fw-190 was flown 29 times between 3 July 1942 and 29 January 1943. It was then partially dismantled and tests done on engine performance at Farnborough. It was struck off charge and scrapped in September 1943.

Faber meanwhile was a prisoner of war in Canada, where he managed to successfully convince British authorities that he suffered from epilepsy. Remarkably, it appears the authorities were taken in by his ruse and in 1944 they allowed his repatriation. Shortly after his return, he was again flying in front-line fighter operations.

After a short period working in America as a test pilot, Wilson joined 616

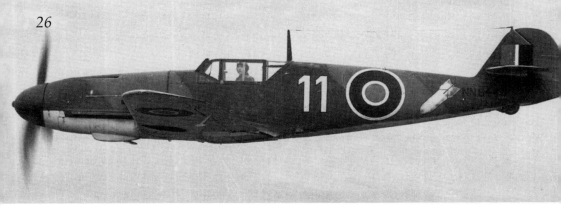

NN644, a Messerschmitt Bf.109F-4 still wearing it's 'White II' and falling bomb symbols going back to its time in service with JG26.

Squadron in 1944 to train pilots on Britain's first jet fighter, the Gloster Meteor. He subsequently became Officer Commanding of the Empire Test Pilots' School at RAF Cranfield.

On 7 November 1945, flying the Gloster Meteor Britannia (EE454) over a 1.86 miles course at Herne Bay in Kent, Wilson averaged a new world air speed record for a jet fighter of 606.38 miles per hour.

After leaving the RAF with the rank of Group Captain, Wilson worked as an engine salesman for Blackburn Aircraft and Rolls-Royce.

Publishing the information
Remarkably – and somewhat surprisingly – throughout the entire period of the Second World War there appeared regular articles in the Allied aeronautical press regarding the activities of the Axis industry and air forces. Often this was very 'up-to-date'.

During the course of the war a number of the aircraft examined at the RAE were subsequently passed to various local and national authorities for exhibition purposes, to raise money for War Bonds and Savings, Spitfire Funds, and other causes.

At the request of the Ministry of Economic Warfare, an extensive and intensive study was undertaken of the construction methods of the German aircraft industry, in order to seek out any weaknesses in the economic and production centres. Therefore, in addition to flying aircraft on test, evaluation of equipment was a primary function, various departments within the RAE being responsible for close study of specific items. These departments comprised Aerodynamics, Armaments, Chemistry, Electrical Engineering, Engines, Instruments, Materials/Metals, Mechanical Test, Photographic, Radio/Wireless and Structures/Airworthiness Standards. As part of this latter item, particular attention was paid to build quality.

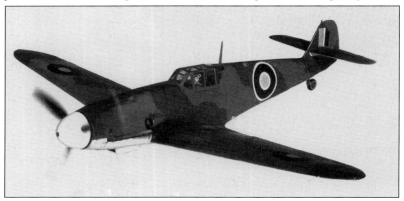

Bf.109G-2/Trop RN228 once 'Black 6' on III/JG77. It was captured at LG139/Gambut Main in Libya on 13 November 1942 and made airworthy by members of 3 Squadron, Royal Australian Air Force. At the controls is Flying Officer G D M Gough.

BEHIND THE LINES

Service and Industrial News from the Inside of Axis and Enemy-occupied Countries

Honours for Dornier

CLAUDIUS DORNIER, founder and managing director of the Dornier Aircraft Works, has been elevated by Hitler's order to the rank of Professor.

Ladies for Regia

ABOUT 700 women, members of the Fascist Party, have been recruited for service with the Italian Air Force. One hundred of them are reported to be engaged in wireless communications from airfields, while the rest are undergoing training as ground wireless operators.

Hispano Suiza

AT their works in Bois-Colombes, near Paris, and at Tarbes, in Unoccupied France, this firm has completed the development of a new 12-cylinder liquid-cooled engine, designated 12Z. The new type differs from the 12Y by the doubling of its valves, and its power output is about 1,400 h.p. It is reported, however, that this will soon be brought up to 1,600 h.p.

The development of a 24-cylinder type demonstrated at the Paris Aircraft Show of 1938 continues under the direction of M. Birkigt. The four 6-cylinder banks in an "H" arrangement and driving two crankshafts are now to be replaced by those of the new 12Z. The new 24 H engine is to have a power output of about 3,000 h.p. A further report suggests that the company is planning the development of a high-performance engine, composed of two tandem arranged 24 H types and developing 6,000 h.p.

V.G.50

THE Vichy Government Aircraft Works at Lyons-Villeurbanne are engaged on the construction of a twin-engine all-metal high-speed aircraft, "V.G.50." The designers are Vernisse and Galtier, who for many years were engaged on the design of wooden aircraft. Of particular interest is the power plant arrangement, which is composed of two Hispano Suiza 12Z engines of about 1,500 take-off h.p. and which drive contra-rotating airscrews. One

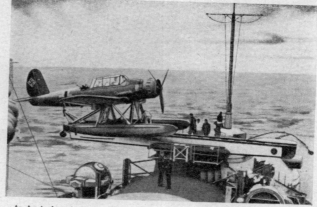

An Arado Ar 196 two-seater reconnaissance aircraft on the retracted catapult of a German cruiser. The Ar 196 is powered with a Bramo Fafnir 323 air-cooled radial of 830 h.p. It will be remembered that they were employed by the Bismarck *during its historic engagement with the Royal Navy.*

engine is mounted in front of the pilot's cockpit, the other in its rear and below it. The extension shaft of the rear engine passes between the legs of the pilot and contains also the pressure oil lines for the operation of the c.p. air-screw.

Two prototypes are now in construction, one with a conventional, the other with a nose-wheel landing gear. The gross weight of the aircraft is about 16,500 lb., wing loading about 41 lb./sq. ft., and wing area between 365 sq. ft. and 387 sq. ft.

It is reported that the designs for a four-engined Atlantic flying boat have been completed by the same firm. The flying boat is to have engines mounted in tandem. (According to "Interavia" a similar arrangement is to be found on the Heinkel He 177.) The aircraft is to be fitted with a pressure cabin; its gross weight is to be about 27 metric tons, span about 138ft., and its speed 310 m.p.h. at 29,500ft.

Nazi Munition Council

FLIGHT, July 9th ("Behind the Lines"), reported the formation of a German Munition Council. As mentioned then, this Council is to co-ordinate military and production requirements. In addition to the military members mentioned in our previous report, the following representatives of the industry have been called by Professor A. Speer, Nazi Munition Minister, to serve on the Council: H. Bucher, managing director of the A.E.G.; P. H. Kessler, chairman of the Bergmann Electrical Works and former director of the Siemens firm, well known for his experience in Japan; P. Pleiger, chairman of the Iron Mining and Smelting Section of the Hermann Goering Works; Dr. E. Poensgen, chairman and managing director of the Vereinigte Stahlwerke (United Steel Works); W. Zangen, chairman and managing director of the Mannesmann Tube Works; H. Roehnert, chairman and man. director of the Rheinmetall-Borsig and chairman of the holding company of the H. Goering Works; Dr. A. Voegler, president of the Kaiser Wilhelm Society.

The above representatives of the munition industry are known as "Leaders of Economy" (Wirtschaftsfuehrer). In addition to this Council, a number of central industrial committees has been formed for the different branches of war production. These committees are composed of engineers and production experts who work in close contact with technical branches of the armed forces and have full authority over the mass production of different war equipment.

The Italian Caproni Reggiane 2001 single-seater fighter. It is powered by a Mercedes-Benz DB601N engine of 1,150 h.p. and has a maximum speed of 348 m.p.h. at 22,000 ft.

Captures abroad

The numbers of Axis aircraft captured in the Middle East and the timing of their capture depended to a large degree on the swings and fortunes of war in the Libyan Desert and elsewhere. At the outbreak of war between Great Britain and Italy, on 10 June 1940 the main strength of the RAF in the Middle East was in Egypt, with smaller contingents in Palestine, the Sudan, Kenya, Aden, Iraq and Gibraltar. Their main strategic purpose was the defence of the Suez Canal. The Italians had nearly three hundred aircraft in Libya, one hundred and fifty in East Africa, and fifty on the island of Rhodes, with a further twelve hundred in Italy itself.

There were several campaigns in the Middle East, not all of which yielded aircraft captures. There is little doubt that the premier collectors and restorers of ex-enemy equipment were the personnel of 3 Squadron, Royal Australian Air Force , which, together with other squadrons of the RAF, RAAF and South African Air Force (SAAF), formed 'Royal Air Force Middle East'. After the USA entered the war, these squadrons were supplemented by units of the USAAF.

The German invasion of Greece and Yugoslavia during April 1941 saw a number of aircraft of German and Italian origin brought into British service. These were Dornier 17K and Savoia Marchetti SM.79K bombers from the Royal Yugoslav Air Force and Dorner Do.22 seaplanes of the Royal Yugoslave Naval Air Service. All aircraft were flown from Yugoslavia via Greece to Egypt.

In East Africa, Italy occupied a number of countries, together with British Somaliland, which had been invaded shortly after Mussolini's declaration of

Messerschmitt Bf 109G-2/Trop, c/n 10639, being inspected before uncrating at Collyweston by Flt Lt E. R. Lewendon, righr, and Flying Officer D. G. M. Gough. The Staffel insignia can be seen aft of the white fuselage band.

What is of particular interest is the amount of labelled 'spares' hanging from the side of the packing case!

This Ju.88D is seen at Heliopolis, Egypt in RAF markings. It eventually ended up with the National Museum of the United States Air Force in Dayton Ohio after being allocated FE No. 1598.

war in June 1940. The Regia Aeronautica deployed one hundred and fifty aircraft in the region.

Between January and April 1941, British forces were on the attack in all these territories. By the end of April 1941, victory was virtually complete. A large number of Italian aircraft were captured, and several were taken into Allied service, including examples of Fiat CR.32, Fiat CR.42, Caproni Ca.133, Caproni Ca.148, Savoia Marchetti SM.73, Savoia Marchetti SM.79 and Savoia Marchetti SM.81.

In Iraq, on 3 April 1941, an Iraqi politician named Rashid Ali seized power in Baghdad. He was an Axis sympathiser and was hostile to the British. By the end of April, tensions had mounted and Iraqi troops had surrounded the main RAF base at Habbaniyah. Under orders to restore its position, the RAF unit at the base, 4 Flying Training School, opened fire on its besiegers at first light on 2 May British troops at Basrah and elsewhere moved to support the RAF and the revolt began to be suppressed.

On 12 May, a Heinkel He.111 landed at Baghdad carrying a German liaison officer to co-ordinate Axis co-operation with Iraq. Next day, an RAF Blenheim was attacked by a Messerschmitt Bf.110, one of a number which had arrived via Syria, which was then controlled by the Vichy French. Other attacks by Bf.110 and He.111 aircraft followed. The German aircraft were based at Mosul in northern Iraq and carried Iraqi markings, but were flown by Luftwaffe pilots. On 28th May, a squadron of Italian CR.42 fighters also arrived in Iraq. By this time it was too late - British forces reached the outskirts of Baghdad on 29 May and an armistice was agreed on 31 May. A small number of German and Italian aircraft were captured.

Following the invasion of North Africa by US and British forces in November 1942, when the US 12th Air Force landed in support of the invasion forces, the combined Allied Air Forces eventually came to be combined as the North West African Air Forces. They continued to support the Allied armies during the later invasions of Sicily, the South of France and the Italian mainland. The end of the war found these air forces (by then renamed the Mediterranean Allied Air Forces) in Northern Italy and Austria, where they became the occupation forces.

By the time of the German surrender in Tunisia, Allied plans for the invasion

Wr.Nr. 263, previously 2N+HT of ZG76 was the first Me.410A to appear in RAF markings when it was 'taken over' by 610 Sqn RAF in Scicily in August 1943.

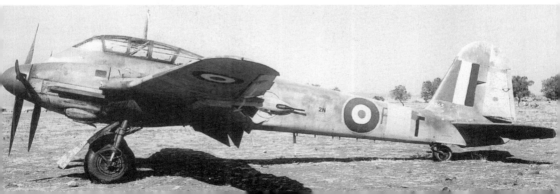

of Sicily had already been made, and the landings started on 10 July 1943. A major objective of the invasion was the capture of the Gerbini group of airfields, inland from Catania, which had originally been built by the Italians to support their planned invasion of Malta, and which now, with Catania airfield itself and Comiso near Vittoria in the south of the island, were the main Luftwaffe fighter bases. Gerbini was not captured until towards the end of the campaign which was concluded on 17 August 1943. Eleven hundred enemy aircraft were found in Sicily, 132 of them at Catania and 189 at Gerbini.

When 3 Squadron, RAAF, landed in Sicily as part of 239 Wing, several more acquisitions appeared. These included a Caproni Ca 100 light aircraft, one of several presented to RAF units by the Catania Aero Club in August 1943. They were generally used for giving air experience to the long-suffering unit ground crews. A number of Messerschmitt Bf.109Gs were also acquired by RAF and USAAF units at Gerbini.

Mainland Italy itself was invaded on 9 September. The main assault was made at Salerno, to the south of Naples, although smaller landings had been made as early as 3 September across the narrow Strait of Messina from Sicily onto the so-called 'toe' of Italy. Ten miles south of Salerno lay the major airfield of Monte Corvino, where several captures were made. Later captures were made at Foggia which fell under Allied control on 27 September 1943.

The deserts of Libya were a happy hunting ground for Axis technology, the three airfields around Gambut being particularly profitable.

Henschel He.129B-1 NF756 was captured in the Middle East and was shipped to the UK in a poor state or repair. It took over a year to get it back to flying condition with 1426 Flight.

Chapter Three

The 'RAFwaffe'

The main specialist service unit within the United Kingdom which operated captured enemy aircraft, expressly to exhibit them to operational units in the British Isles, to give ground and flying demonstrations for the benefit of ground staff, aircrew and aerodrome defence personnel, and for fighter liaison and aircraft recognition duties, was formed at RAF Duxford on 21 November, 1941, and was called 1426 (Enemy Aircraft) Flight.

Initially the flights were to RAF and later USAAF 8th and 9th Air Force bomber airfields, but the demonstrations were extended to cover ground defence organisations such as anti-aircraft artillery units and personnel of the Royal Observer Corps.

This organisation maintained a network of ground-based observation posts all over the United Kingdom, each linked by telephone to the fighter and other anti-aircraft defences. Accurate aircraft recognition was an essential skill of the ROC observers and much effort was put into their training. While much of this was theoretical classroom study - or study of books at home, for almost all Observer Corps personnel were people with civilian jobs working as Observers in their spare time - flying displays of aircraft were held from time to time as part of their ongoing training.

The initial personnel consisted of Flying Officer R. F. Forbes, the first Commanding Officer, Flying Officer Kinder, Pilot Officer E. R. Lewendon and Flight Sergeant D. G. M. Gough, all being posted from the Air Fighting Development Unit (AFDU) - also at Duxford - to which they had been attached for eleven days for flying experience on German aircraft. The group also visited Farnborough to fly those aircraft that were based there prior to being transferred to the flight. All the pilots had previously been in 41 Group as Maintenance Unit test pilots. The Flight came under the operational control of 12 Group, Fighter Command.

The selection of test pilots to staff the Flight showed considerable forethought on the part of the planners. Unit records show that almost all flights made by the German aircraft were of the nature of test flights and that many of them ended prematurely due to some emergency or a prudent

Bf.109E DG200, seen in flight after its canopy had somehow been 'lost'. This machine had been coded 'Black 12' of I/JG51 and had been forced-landed at Manston in Kent on 27 November 1940. *(Peter H T Green Collection)*

Junkers Ju.88R PJ876 seen airborne during another fighter affiliation sortie. By now the nose radar aerials had been removed, and the 'prototype P' marking had been applied in yellow. (Peter H T Green Collection)

return to base because of a potential or actual problem. The 'tours' of other flying stations included a constant catalogue of cancelled demonstrations and shuttles in communications aircraft to the Flight's base or to RAE Farnborough to obtain spares or tools needed to get one or other of the aircraft back into the air.

Duxford airfield had came into being during the latter stages of World War One, when in October 1917 P & W Anderson Ltd began construction of the airfield at an estimated cost of £90,000. All materials were brought to the site from Whittlesford railway station by either steam lorries or horse-drawn wagons.

As the work went on, more and more men were employed; at one stage it was estimated that a thousand men were working on the project. The cost of the work also began to rise, but progress was slow.

In March 1918, United States Army personnel of the 159th Aero

Aircraft on Strength of 1426 Enemy Aircraft Flight.

Serial	Type	Dates
AE479	Messerschmitt Bf.109E	11 Dec 1941 to 28 Jan 1942 (shipped to USA)
AW177	Heinkel He.111H	7 Dec 1941 to 10 Nov 1943 (lost in accident)
AX772	Messerschmitt Bf.110C	5 Mar 1942 to 31 Jan 1945
DG200	Messerschmitt Bf.109E	28 Apr 1942 to about September 1943
EE205	Junkers Ju.88A	28 Aug 1942 to 31 Jan 1945
HM509	Junkers Ju.88A	11 Dec 1941 to 19 May 1944 (lost in accident)
NF754	Focke Wulf Fw.190A	12 Dec 1943 to 31 Jan 1945
NF755	Focke Wulf Fw.190A	12 Dec 1943 to 31 Jan 1945
NF756	Henschel Hs.129B	27 Jun 1943 to 31 Jan 1945
NN644	Messerschmitt Bf.109F	21 Aug 1943 to 31 Jan 1945
PE882	Focke Wulf Fw.190A	19 Apr 1944 to 13 Oct 1944 (lost in accident)
PJS76	Junkers Ju.88R	6 May 1944 to 31 Jan 1945
PN999	Focke Wulf Fw.190A	28 Sep 1943 to 31 Jan 1945
RN228	Messerschmitt Bf.109G	26 Dec 1943 to 31 Jan 1945
TS472	Junkers Ju.88S	25 Sep 1944 to 31 Jan 1945
VD364	Messerschmitt Bf.109G	(arrived 14 February 1945, as the unit disbanded)
VX101	Messerschmitt Bf.109G	9 Apr 1944 to 19 May 1944 (lost in accident)

Heinkel 111 AW117 is seen in flight over the English countryside. *(Peter H T Green Collection)*

Squadron arrived to erect Bessonneaux hangars as a temporary measure to protect aircraft while the main hangars were completed. The permanent hangars would consist of three double-bay and one single-bay Belfast types, with wooden concertina-type doors at each end; these buildings took the name from the wooden roof trusses they incorporated.

As soon as the Bessonneaux hangars had been completed, the airfield was used as a mobilisation station, and three squadrons, 119, 123 and 129 arrived in March and April 1918 with their DH.9 bomber aircraft. More Americans also arrived in the shape of the 151st, 256th and 268th Aero Squadrons, but most of their work was in assembling aircraft and running the Motor Transport section. 129 Squadron, now part of the newly-formed Royal Air Force, disbanded in July 1918, and that month a Training Depot Station opened at Duxford as part of a new system of RAF pilot training. RAF Duxford was not opened until September 1918, and even then the contractors were still toiling on, as slowly as ever! By the end of the War in November 1918, 35 TDS was made up of 450 men and 158 women, the latter engaged in clerical and domestic work.

The Americans packed and left very soon after Armistice Day and things became quiet at Duxford. When war ended, several airfields found themselves without a job. Duxford's future was questioned and plans were revised, after which it was decided to keep Duxford as a permanent RAF Station, to be used for flying training.

Throughout the inter-war years activities at Duxford was fairly low key,

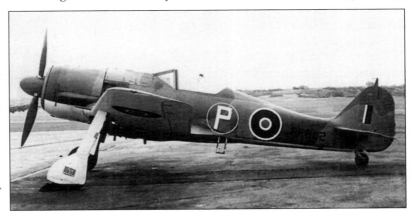

Focke Wulf 190A wearing the RAF serial PE882. The 'P' in the circle in yellow was usually used on prototype aircraft. The aircraft, formally in the possession of II/SKG10, orginally came into the RAF's possession after it landed by mistake at West Malling on 17 April 1943. *(Peter H T Green Collection)*

until 4 August 1938, when an event took place which ensured Duxford a place in the history books. Technology was moving forward on all fronts, and the development of aircraft had taken a mighty leap forward when R. J. Mitchell designed the Spitfire and on that afternoon Jeffery Quill flew Spitfire K9789 into Duxford for delivery to 19 Squadron.

A steady stream of Spitfires continued to arrive at Duxford, and 66 Sqn also began to convert onto the type. Gauntlets, Gladiators and other outdated aircraft began to fly into history, and the Spitfire reigned supreme at Duxford! Training on the new aircraft continued into 1939. Various problems were overcome, although there was the usual crop of accidents which happened when new types were introduced, but the pilots and ground crews liked what they had been given.

The summer wore on and tension mounted again. 611 (West Lancashire) Squadron of the Royal Auxiliary Air Force arrived early in August for summer Camp, bringing its Spitfires. All three squadrons went to half-hour readiness on 24 August, and personnel away from the Station were recalled. Perimeter security was stepped up, and on 29 August all reservists were called up.

Emergency procedures were once again implemented, including the sand-bagging of various installations around the airfield. 1 September, the day that Hitler's forces invaded Poland, saw the general mobilisation of the RAF, and two days later, on Sunday 3 September 1939, Britain and France declared war on Germany. The Station records show that *'...Signal A34 received, announcing war has broken out with GERMANY only'*

Security became even tighter, and work began on modification of barracks and other buildings in readiness for WAAF personnel. A false air raid alarm on 6 September kept everyone on their toes and flying movements became hectic. Camouflage paint was applied to the hangars, no civilian clothing was allowed on the Station and air raids or an invasion, or both, were expected at any minute. Neither happened, and life became routine again, but an air of expectancy prevailed.

The airfield was to play a vital part in the Battle of Britain, with possibly the most famous of all pilots being based there - Douglas Bader. The end of the Battle, with the thwarting of Hitler's invasion plans, saw things change.

At the end of 1940, Duxford was set to change its role again. The Air Fighting Development Unit had been formed to evaluate new aircraft and systems alongside the Air Gun Mounting Unit, which concentrated on

Focke-Wulf Fw.190A PN999 seen under the wing of Junkers Ju.88S-1 TS472 at Collyweston as recorded by a photographer of the USAAF. *(USAAF Photo).*

Messerschmitt 109E-3 serialled AE479 seen airborne from Duxford in 1941. This aircraft was shipped to the USA for further evaluation in 1942. (USAAF Photo)

armament. Some of the work on the cannon-armed Spitfires had been done by this organisation. Both units were starting to move into Duxford in late 1940 and by the spring of 1941 had settled into a varied routine of work. Alongside them was the Naval Air Fighting Development Unit, 787 Sqn, which flew shipboard types such as the Fulmar and Martlet.

With this form of development flying taking place on the airfield, it was not surprising that the formation of the enemy aircraft flight was decided upon. The first aircraft allotted was Heinkel He.111H-1 AW177 from the RAE on 7 December, 1941, followed by Messerschmitt Bf.109E-3 AE479 and Junkers Ju.88A-5 HM509 on 11 December 1941. The posting of maintenance personnel began on 22 December, 1941, and some of the ground crews were sent to the RAE for instruction in the maintenance of German aircraft. An addition to the Flight on 2 February 1942, was Flight Sergeant F. A. Barr, who was posted to the unit for full flying duties.

The Tours
The first tour - No. 1 tour - of RAF stations began on 11 February, 1942, using two aircraft; He.111 AW177 and Ju.88 HM509. They flew with an escort of Spitfires from Duxford to Lakenheath, overflying en route the airfields of Oakington, Upwood, Marham, West Raynham, Swanton Morley and Mildenhall. At Lakenheath the aircraft were inspected by members of the Royal Observer Corps.

On 13 February the pair flew to RAF Watton, giving further displays at Lakenheath and Mildenhall. Another was given at Watton on 14 February and on the following day the aircraft flew to Coltishall and gave two flying displays there on 16 February. Next day both aircraft flew to Bircham Newton, where the display included a dogfight between the Ju.88 and one of the locally-based Lockheed Hudsons. 18 February saw the pair fly to nearby Docking and perform dogfights with a Beaufighter before returning to Bircham Newton. On 19 February the aircraft flew to Sutton Bridge and on the next day carried out mock dogfights with the Hurricanes of the Sutton Bridge-based 56 Operational Training Unit. The Flight moved over to Wittering on 28 February and next day the Ju.88 made two demonstration flights but the Heinkel remained on the ground because of a propeller problem. Finally, on 27 February, both machines returned to Duxford.

On 5 March 1942, Messerschmitt Bf.10C-5 AX772 was posted in from the Royal Navy AFDU, just before No. 2 tour started on 1 April 1942. Demonstrations were given at RAF stations North Luffenham, Cottesmore, Saltby, Cranwell, Digby, Waddington, Hemswell, Kirton-in-Lindsey, North

An unidentifiable Bf.109G on what is clearly an American airfield. It is thought that it could be RN228. *(Peter H T Green Collection)*

Coates and Snaith. On 28 April, 1942, Messerschmitt Bf.109E-3 DG200 was delivered by road in a dismantled condition. No. 3 tour began on 4 May, 1942, and covered RAF stations Church Fenton, Holme-on-Spalding Moor, Breighton, Leconsfield, Catfoss, Driffield, Pocklington, Marston Moor, Linton-upon-Ouse and Dishforth. No. 4 tour followed on 15 June 1942, to RAF stations Cranfield, Bassingbourn, Twinwood Farm, Waterbeach, Stradishall, Newmarket, Wattisham, Debden, Castle Camps, North Weald, Hornchurch, Bradwell Bay, Fairlop, West Malling, Detling and returned to Duxford on 3 July, 1942.

So was established the flying routine. Occasional local flights were made from Duxford to the nearby Army training areas to give demonstrations. A number of flights were made quite frequently for the benefit of the RAF Film Unit or other organizations involved in making recognition films or sound recordings of enemy aircraft types.

Meanwhile, on 12 June 1942, Junkers Ju.88A-ls AX919 and HX360 arrived by road from the RAE and were used as a source of spares to service HM509 and EE205. On 25 July 1942, No. 5 tour, comprising a Bf.109, Ju.88 and He.111, visited Biggin Hill, Gravesend, Kenley, Redhill, Atcham, Heston, Northolt, Aircraft and Armament Experimental Establishment (A&AEE) Boscombe Down, where a number of Establishment test pilots flew the aircraft for familiarisation, and the Ju.88 HM509 was flown at night for exhaust glare photographs of its flame-damping exhausts. The tour continued to USAAF stations Bovingdon and Molesworth.

The Junkers Ju.88A-5 EE205 was flown in from the RAE on 28 August, 1942, and No. 6 tour began on 4 September to USAAF station Atcham and RAF stations Ibsley, Old Sarum, Andover and Colerne. Continuing the 'merry round', No. 7 tour on 14 October 1942 covered Charmy Down, Weston Zoyland, Church Stanton, Exeter, Harrowbeer, Chivenor, Netheravon, Hullavington and Castle Combe. No. 8 tour, on 30 November, 1942, visited Aston Down, Chedworth, South Cerney, Abingdon, Bicester, Upper Heyford, Chipping Warden and Bovingdon. During January 1943 Bf.110 AX772 was flown on ground strafing exercises for the RAF Regiment. On 24 February 1943, No. 9 tour began and visits were made to USAAF stations Debden, Castle Camps, Bassingbourn, Thurleigh, Chelveston, Molesworth, Alconbury, Honington, Hardwick, Bungay, Shipdham, Horsham St Faith, Swanton Morley, Foulsham, West Raynham and Great Massingham, returning to Duxford on 9 April 1943, the unit having moved

Two views of Messerschmitt Me.110C AX772 seen during one of the 1426 Enemy Aircraft Flight 'tours'. The machine served for a considerable time with the RAE before passing to the EAF and was transferred to Sealand for storage in 1946. It is believed to have been scrapped sometime during 1947 or 1948. (USAAF Photos).

to RAF Collyweston on 12 March 1943. On 27 May the Flight went to RAF Digby, then occupied by Canadians, where they were inspected by King George VI and Queen Elizabeth.

In June the first aircraft to be captured in the Middle East arrived, a Macchi MC 202 that was almost complete. Two technicians from the Enemy Aircraft Section at Farnborough arrived to report on the aircraft, but it was suffering too badly from corrosion to be rebuilt.

Henschel Hs.129B-1 NF756 arrived in packing cases on 27 June and a party of airmen from 65 MU Blaby was attached to the Flight to assist in the rebuild. On 7 July He.111H-1AW177 and one of the Ju.88s was flown to USAAF station Polebrook to enable Captain Clark Gable to make an instructional film for air gunners. On 10 August some members of the Enemy Aircraft Non-Ferrous Materials Committee and representatives of High Duty Alloys visited the Flight to study the materials and construction of the Hs.129. Messerschmitt Bf.109F-4/B NN644 was received, via the RAE, on 21 August followed on 28 September by Focke-Wulf Fw.190A-4/U8 PN999, also from RAE.

Late in September a group of Polish pilots was attached to the Flight to enable them to familiarise themselves with the aircraft. The senior Polish pilots, Squadron Leaders Bialy and Iszkowski, were allowed to fly the He.111, Ju.88 and Bf.110.

On 5 November Bf.109F-4/B NN644 was flown for air-to-air

Two views of Junkers Ju.88A HM509 as it is inspected by a number of American servicemen on a wet airfield 'somewhere in England'. Originally from Küstenfliegergruppe 106 and serialled 'M2-MK', this aircraft left Morlaix at 16.00 on 26 November 1941, to attack shipping which had been reported in the Irish Sea. It flew up the Irish Sea to a point between Bardsey Island and Wicklow but in spite of a prolonged search no shipping was sighted. Fuel began to run low and the crew turned for home. Due to a miscalculation the crew became lost and eventually landed at RAF Chivenor, thinking they were in France. Slight damage occurred during its capture when a Tommy Gun was fired at the cockpit, but this was soon repaired. *(both USAAF)*

photography in company with a Hudson, flown by Flt Capt James Mollison of the Air Transport Auxiliary.

No. 10 tour began on 6 November, 1943, consisting of He 111H-1 AW177, Ju.88A-5 HM509 and Bf.109F-4/B NN644, visiting USAAF stations Goxhill, Grafton Underwood and arriving at Polebrook on 10 November. That day the Flight ended a demonstration at Goxhill at 17.30 hours and flew to Polebrook, led by the He.111 flown by Flying Officer Barr. On arrival at Polebrook he put the wheels of the He.111 down, followed by the Ju.88. The '88 turned in to land on the illuminated duty runway. The '111 continued its circuit and started to approach the same runway from the upwind end. Its pilot saw the Ju.88 landing and opened up to avoid him, making a steep left hand turn. The aircraft spun in from 100 feet, hitting the ground vertically. Seven of the passengers were killed by the impact when the fuel tanks exploded, but four others escaped with injuries. The Bf.110, which was following the He.111, and a Bf.109 following the Ju.88, both landed safely.

On 12 December, 1943, two Fw.190s, NF754 and NF755, arrived, and on 26 December Messerschmitt Bf 109G-2/Trop RN228 followed, all in packing cases via the Middle East and Liverpool Docks. No. 11 tour began on 31 December, 1943, and visited USAAF stations Molesworth, Chelveston, Kimbolton, Thurleigh, Poddington, Bassingbourn, Steeple Morden and Cheddington, returning to Collyweston on 1 February, 1944.

That same month the Unit took on additional duties, with the Flight's Bf.109G, flown by F/Lt E. R. Lewendon, being used for comparative performance trials with a Hawker Tempest I of the AFDU. Later in the month, further trials were carried out against a Mustang III and a Spitfire XIV, also of the AFDU. In March, this series of tests continued against a Seafire III, F4U Corsair and Hellcat of the Naval AFDU, followed by tests of the Hellcat and Seafire against an Fw.190.

Messerschmitt Bf.109G-6/Trop VX101 arrived on 4 February, and on 23 March No. 12 tour began to RAF and USAAF stations Hullavington, Bovingdon, Chipping Ongar, Stansted, Great Dunmow, Great Salling, Earls Colne, Rivenhall, Ridgewell, Wattisham, Boxted, Raydon, Martlesham Heath

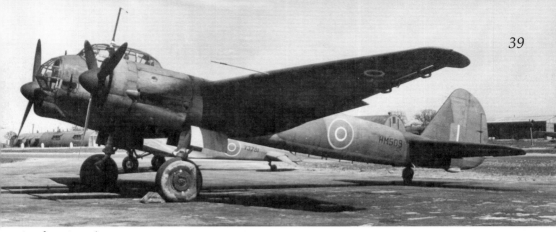

Another view of Junkers Ju.88A HM509

Messerschmitt Bf.109F-4 (RAF serial NN644) parked near the control tower at Bassingbourn, Cambridgeshire, during the unit's 11th tour of operational stations giving flying demonstrations. Although painted in RAF colours, the aircraft retains the 'White 11' and bomb symbol markings of its former Luftwaffe unit, 10.(Jabo)/JG 26. *(Vince Hemmings Collection)*

and Framlingham, returning to Collyweston on 5 May.

Focke-Wulf Fw.190A-4/U8 PE882 arrived from the RAE on 23 April, 1944, followed on 6 May by the Junkers Ju.88R-1 PJ876. On 9 May the Flight moved to RAF Thorney Island, for five weeks' attachment, for recognition exercises over the invasion fleet on the south coast to form part of an 'Air Show Circus' with eighteen British and American aircraft types. The purpose was to fly over various Allied units during the build-up to D-Day to provide instruction in aircraft recognition. On 9 May, Ju.88 HM509, Fw.190s PE882 and PN999, and Bf.109 RN228 positioned to Thorney Island, to be joined on 26 May by Bf.110 AX772 and Ju.88 PJ876. Flights continued until the invasion on 6 June enabled a return to normal duties.

The Flight began a special recognition exercise for the Mosquito squadrons based at RAF West Raynham and Little Snoring on 9 August, 1944. During the preceding month, on 11 July, the Focke-Wulf Fw.190A-5/U8 PM 679 arrived by road from the AFDU Wittering, to be used as a source of spares to service PN999. On 18 September, 1944, the Fw.190 PN999, Bf110 AX772 and Ju.88 EE205 were flown to USAAF Chipping Ongar, where they were used for instructional purposes by the USAAF Disarmament School, for the maintenance and temporary immobilisation of German aircraft.

A few days later, on 25 September 1944, Junkers Ju.88S-1 TS472 arrived from Villacoublay, and on 27 September Flt Lt R. F. Forbes was posted from the Flight to 61 OTU. Flt Lt E. R. Lewendon had succeeded him as CO that August.

Sixteen Fw.190 are known to have been evaluated in the UK. The first was captured on 23 June 1942, when Lt Arnim Faber of III/JG2 became disorientated and landed his Fw.190A-3 at RAF Pembrey in south Wales. His aircraft went to the Air Fighting Development Unit (AFDU) at Duxford for

Right: Heinkel He.111H-1, c/n 6853, carried the badge of the Löwen (Lion) *Kampfgeschwader*, KG26, on both sides of the fuselage, forward of the wing leading edge. This emblem was retained while in British markings.

Below: A view inside the bomb-bay, looking aft.

Bottom: AW177 undergoes inspection.

comparative trials against Allied fighter types, including the Spitfire VB, which then equipped most Fighter Command squadrons. The following excerpts are from AFDU documents, which must have made grim reading at the time for the RAF: *'The Fw.190 was compared with a Spitfire VB from an operational squadron for speed and all-round manoeuvrability at heights up to*

Left to right: Flight Sergeant Lee, Flying Officer Staples, Flying Officer D G M Gough, and Flight Lieutenant E R Lewendon at RAF Collyweston, with Bf.110 AX772 in the background.

25,000ft. The Fw.190 is superior in speed at all heights, and the approximate differences are listed:

- At 1,000ft the Fw.190 is 25-30 mph faster than the Spitfire VB.
- At 3,000ft the Fw.190 is 30-35 mph faster.
- At 5,000ft the Fw.190 is 25 mph faster.
- At 9,000ft the Fw.190 is 25-30 mph faster.
- At 15,000ft-18,000ft the Fw.190 is 20 mph faster.
- At 21,000ft the Fw.190 is 25 mph faster.
- At 25,000ft the Fw.190 is 20-35 mph faster.

The climb of the Fw.190 is superior to that of the Spitfire VB at all heights. Under maximum continuous climbing conditions the climb of the Fw.190 is about 450ft/min better up to 25,000ft. With both aircraft flying at high cruising speed and then pulling up into a climb, the superior climb of the Fw.190 is even more marked. When both aircraft are pulled up into a climb from a dive, the Fw.190 draws away very rapidly and the pilot of the Spitfire has no hope of catching it.

Comparative dives between the two aircraft have shown that the Fw.190 can leave the Spitfire with ease, particularly during the initial stages.

The manoeuvrability of the Fw.190 is better than that of the Spitfire VB except

Aircraft from 1426 (Enemy Aircraft) Flight visit Goxhill, one of the many USAAF bases in England.

Focke-Wulf 190A-4/U8, formerly 'Red 9' of I/SKG 10. This machine landed by mistake at Manston in Kent in the early hours of 20 May 1943. It was later flown by RAE as PN999 and was used by 1426 (Enemy Aircraft) Flight. *(Peter H T Green Collection)*

in turning circles, when the Spitfire can quite easily out-turn it.
The Fw.190 has better acceleration under all conditions of flight and this must obviously be useful during combat.

With Operation Overlord, the invasion of Europe, starting in June 1944, it is not surprising that the number of captured aircraft increased. One such example was Heinkel He 177A-5/R6 W Nr 550062. This He.177, coded 'F8+AP' of II/KG 40, was captured by the French Resistance at Toulouse-Blagnac where it was under overhaul in the Ateliers Industriel de l'Air facility in September 1944. It also had the radio call-sign 'HM+UK' and wore the number '60' in yellow on its fin. It was flown from Toulouse to Farnborough by Wing Commander Falk on 10 September. It wore French markings, including the title *'Prise de Guerre'* applied by the French Resistance. By 20 September it had received RAF markings and the serial TS439. Its first flight was made by Squadron Leader A F Martindale on that day. It was flown on a number of occasions at Farnborough, its last recorded flight being 20 February 1945, when it landed at Boscombe Down. From here it is believed to have been shipped in dismantled condition to the USA where it became 'FE-2100' on arrival. On its last flight the '177 sustained irreparable damage to one of its DB610 engines but this was not a problem as the Americans had obtained a stock of spare engines. (The USAAF had received another He177A in Europe, but this was destroyed at Paris/Orly in a take-off accident at the start of its delivery flight to the USA).

TS439 made twenty flights under RAE control (including its ferry flight), totalling 19 hours 35 minutes. Trials covered aspects as varied as tests of the spring tab controls, bombsight and altimeter tests and checks of the crew's air heating system. Pilots who flew the aircraft included Squadron Leader Weightman, F/Lt Turner, Squadron Leader Keeling, F/Lt Wellwood, Squadron Leader Randrup and Group Captain A. F. Hards.

On 13 October, 1944, while doing local flying in Fw 190A-4/U8 PE882, Lewendon was killed when the aircraft crashed near Collyweston. With effect from 14 October, Flt Lt D. G. M. Gough assumed command of the unit. On 5 December, 1944 Ju.88S-1 TS472 was flown on an affiliation exercise with a Lancaster from RAF Woolfox Lodge, followed on 18 December by

Heinkel 177A-5/R6 'Grief' as flown by Wing Commander 'Roly' Falk on 10 September 1944 after being capture by the French Resistance at Blagnac, France.

recognition exercises at RAF Ossington; also on 14 January, 1945, a Ju.88 was flown to RAF Kenley, by Flight Sergeant Bennett, for exercises on the temporary immobilisation and maintenance of German aircraft. On 19 January 1945, a working party was flown to Deurne (Antwerp) to collect two Messerschmitt Bf.109G-14/U4s, VD358 and VD364. On 21 January official notification was received of the disbandment of No. 1426 (EAC) Flight at Collyweston with effect from 17 January 1945, reforming at RAF Tangmere on the same date, with unit codes EA, as the Enemy Aircraft Flight of the Central Fighter Establishment, which finally disbanded 31 December 1945. During its existence it had met with many maintenance difficulties, due to the lack of spares and maintenance details. Tools and equipment had to be specially made, and all engine and airframe spares obtained from crashed and unserviceable aircraft. It was also necessary to assemble aircraft that had never been in the United Kingdom and about which little was known from a maintenance point of view.

Evaluating a Gustav

Roland Beamont - later of English Electric Lightning and TSR-2 fame - flew Messerschmitt Bf109G-6 RAF serial VD364 from the Enemy Aircraft Flight at the Central Fighter Establishment's Tactics Branch at Tangmere in the summer of 1945. *'Since first having seen Bf.109E 'Emils' in France in 1940, and having had a close look at one in combat over Lyme Bay in August 1940 just before it belly-landed on fire at Abbotsbury, and another - a 'Gustav' - over Rouen in 1944, I had wondered how the Luftwaffe could cope with what seemed to be very restricted vision from the cockpit.*

Gustav VD364 still retained the original windscreen and side-hinged canopy arrangement, and as I settled into the narrow cockpit - which fitted tightly at the

Bf.109E AE479 - formerly 'White I' of JG76, seen at Duxford with the AFDU sometime early in 1941. The machine had a varied test career, being crated and shipped to the USA in April 1942. It was assigned to Wright Field OH, from where it was destroyed in a forced-landing at Cambridge OH on 3 November 1942. (Peter H T Green Collection)

shoulders - and closed and locked the canopy, there was an immediate feeling of claustrophobia - my first impression had been correct! Vision out of this small aeroplane was difficult even for taxying.

The cockpit layout was primitive after the tailored 190, and showed evidence of its production development through many sub-series with an abundance of add-on instruments and brackets.

The Daimler-Benz engine was rough and raucous after the 190's BMW 801, and taxying required a weaving path similar to that on all Spitfires, in order to see round the long nose.

A gusting south-westerly wind caused the 109 to rock gently on its narrow undercarriage, and frequent use of the full-back-stick tailwheel lock was needed to stop incipient swinging. The 109 felt touchy from the start.

Following the briefing notes carefully I increased power progressively during the take-off, easing the stick forward to raise the tail at about 60 mph and countering swing with the rudder which became effective at quite low speed.

The G-6 became airborne at about 70 mph with only slight aft stick and immediately gave the impression of being a hot little ship. The big engine thundered in front, the ailerons were light and delicate on the climb at 165 m.p.h., and the all-round vision was severely restricted by the heavy windscreen and armour glass frames in front with Revi gunsight in the middle, and by the sideways and upwards view through small square transparencies in the metal-framed hood; the vital rearwards vision was virtually non-existent.

All this gave a first impression of rushing along in a tunnel, and it was only when checking the ASI and altimeter that the performance was actually seen to be less than impressive.

Up to 25,000ft (there was a British ASI and altimeter fitted) level performance was at least 50 mph down on the Tempest V2 but at 30,000ft and above the gap was closed and the G-6 clearly had superior climb performance above about 15,000ft. But there its advantages ended.

In turning manoeuvres it was similar to the Tempest up to 20,000ft; but in rate of roll it could only match the Tempest up to about 350 mph IAS. The ailerons stiffened up rapidly and at anything over 450 mph they were virtually solid! This was no fighter to match the 545 mph IAS manoeuvrability of the Tempest, or even the 485 mph limit of the Mustang.

As expected with its slim, high fineness-ratio fuselage and adequate fin and rudder area, directional damping was crisp and it had the makings of a good gun-platform. Combat manoeuvres in the 250-350 mph range were good, but above that

Messerschmitt Bf-109E-3 'White 4' once operated by 4/JG26 based at Marquise-Est was captured after a forced landing in East Dean, Sussex on 30 September 1940 virtually undamaged. It was sent to Canada and the USA, and is seen being unloaded at New York in June 1941 during a 'Bundles for Britain' campaign.

RAF Messerschmitt Bf.109 serial number NN644 with a B-17 Flying Fortress of the 379th BG at Kimbolton on 8 January 1944. (Peter H T Green Collection)

the heavy ailerons were soon severely limiting, and in hard turning below 200 mph the stall boundary was approached with heavy and disturbing wing rocking caused by asymmetric leading edge slat operation.

As I winged over back down into the Tangmere circuit, I felt that this aeroplane had little to commend it by comparison with its Allied contemporaries, with the exception of its famous fuel injection system which gave it superior negative-g capability.

Setting up a standard curved 'Spitfire approach' seemed quite natural in this long-nosed fighter. The undercarriage and flaps went down without exceptional trim change, and the final turn to wings-level over the threshold in a light cross-wind from port felt tight and secure, although the fact that the side-hinged canopy could not be slid back in the fashion of the time was irksome. Then came the complication.

As I flared and throttled back at about 85 m.p.h., the crosswind made itself felt and I instinctively decided on a 'wheeler', keeping the tail level until the mainwheels touched before beginning to lower the tailwheel. But the Gustav was not about to settle for this and, despite countering with aileron and rudder, the crosswind caused the starboard mainwheel to touch first with a squeal, and this resulted in a gentle series of bounces which was eventually ironed out tail-high until the speed decayed and I lowered the tailwheel into contact, whereupon immediate tailwheel lock was needed to keep straight.

This aeroplane was unforgiving, so I called the tower for clearance, took off and tried it again - with similar results!

In no way would this aircraft accept either wheeler or three-point landings on a runway with a 10kt crosswind without protest, and it was clear that grass field operation permitting take-offs and landings directly into wind would be an

A captured, but unidentified Bf.109G pictured at Gerbini 3, Scicily on 5 September 1943. The machines true identity has been covered by a coat of paint and the RAF roundels.

operational necessity if landing accidents were to be held to a minimum. This was, of course, the experience of 109 operators throughout the war years, in which the landing accident rate had been seriously high.

I taxied the Gustav back to dispersal with wings rocking delicately on its narrow undercarriage, and weaving it from side to side for forward vision amid gusts of exhaust smoke. It was not a good fighter by comparison with the Allies best, but a capable one in the hands of a good pilot and one which I supposed could have found favour with pilots who had not experienced anything better. But I, at least, would not have enjoyed going to war in a fighter with that awful cockpit vision.

One area where a number of German aircraft were captured was the North African desert, as evidenced by this Messerschmitt 110.

Test Reports

The reports of how captured aircraft were examined are many, incredibly detailed, and were usually drawn up very quickly. To tell of each report would fill many volumes - however, some are much more historically important than others. One such example is this report drawn up by Squadron Leader H. F. King from Air Intelligence 2(g). This was the incident when a Junkers Ju.88 G-1 Night Fighter 4R-UR based at Volkel in the Netherlands being flown by a crew of 7/NJG 2 landed at Woodbridge, Suffolk, after flying a reciprocal compass course in error in the early morning of 13 July 1944. Crew: Unteroffizier Hans Mackle, Obgefr. Heinz Olze and Obgefr. Hans Mockle.

Inspection of Crashed or Captured Enemy Aircraft
Report Serial No. 242 dated 16th July 1944
Report No. 8 / 151: Junkers Ju.88 G-1 Night Fighter

Security: In consultation with Air Directorate of Intelligence (Science) it has been decided to grade this report as Top Secret owing to the very special nature of the Radar equipment carried by the aircraft described.

Introduction: At 04.25 hours on 13th July 1944, a Junkers Ju.88 G-1 made a wheels-down landing on Woodbridge emergency landing strip. The pilot was completely lost and had apparently been flying on a reciprocal course to that

The Enemy Aircraft Flight visits North Weald sometime during 1943.

intended. When he sighted Woodbridge and believed himself to be near Berlin, and being very nearly out of petrol he decided to land immediately. He had, in fact, so little fuel and oil, that it was impossible, subsequently, to obtain any samples for analysis.

This important capture is one of Germany's latest night fighters and is fully equipped with up to date equipment.

Radar and radio.
The airframe is essentially that of a Ju.88 C or R, except that a Ju 188 tail-unit had been fitted. Apart from this the main difference between the 'G' and former sub-types consist of improved armament and increased armour protection. The minor differences are numerous and the more important of these will be discussed under the appropriate headings below. The Ju.88 G-1 aircraft has now been flown to the Royal Air Force for flying trials and radar tests.

General Description
Identification: Werk Nummer: 712273
Unit codes, Staffel and aircraft identification: 4R Ë UR
Radio Call-sign (Stammkennzeichen): GF Ë XO

The werk nummer appeared not only on the fin, but on all control surfaces and detachable panels.

The camouflage is Duck-Egg Blue on all surfaces, with Dark-Grey mottling superimposed on the top surfaces.

Engines: Two B.M.W. 801 G-2 radial-engines, fitted with V.D.M. three-bladed propellers. Port engine Nr. 326729; starboard engine Nr. 327293. The manufacturers are unknown at this moment as it is not possible to obtain the data plates without considerable dismantling.

Armament: The forward-firing armament comprises four fixed Mauser MG 151 cannons of 20-mm. calibre. These guns are mounted in a large blister, measuring 11 feet 3 inches x 12 inches deep (maximum) x 26 inches wide, on the port underside of the fuselage. The barrels of the upper guns protrude 28 inches, those of the lower pair 6 inches. The breeches are in the forward bomb-bay, and above them are the ammunition tanks which are estimated to hold 250 rounds each.

None of the guns had been fired, and it is of particular interest that the total

The RAF's Enemy Aircraft Flight puts on a display, with a Bf.110 overflying a Lysander, with a Dornier Do.17 banking in the distance. *(Peter H T Green Collection)*

Ju.88 TP190 in flight.

number of rounds was only 545. Two tanks had 146 rounds each, the third had 145 and the fourth 108.

On the instrument panel, straight in front of the pilot, is a rounds counter suitable to six fixed guns, the last two sections being left blank.

The belted loading-order of these guns was: one Armour-Piercing/Incendiary, two H.E./Incendiary-Tracer, and two Incendiary/Tracer (night-trace or 'glimmer'). A Zeiss Revi 16.D gun-sight is used in conjunction with the forward-firing armament. Externally, this gun-sight differs in two main essentials from the Zeiss Revi C.12/D, which has hitherto been found in German fighters, these are (1) the control for elevating the sighting screen has been modified, (2) the coloured antiglare shield has been removed. If further examination discloses any changes in the sighting mechanism, a full report will be issued to the departments concerned.

The rearward firing armament consists of a single Rheinmetall-Borsig MG 131 of 13-mm. calibre in a manually operate dorsal ring. 235 rounds were carried for this gun in a tank designed for 500, the belted loading order being: one A.P./Tracer (night trace) and one H.E./Incendiary (night trace).

The armament previously found in Ju.88 C and R night fighters examined in this country consisted of: three x MG-FF Oerlikons of 20-mm. calibre and 3 x MG 17 of 7.9-mm. calibre, Fixed, and forward-firing in the nose; and 1 x MG 131 in a manually operated dorsal-ring rearward firing.

Bombing Installation: There was no bombing equipment on this aircraft.

Armour: The entire windscreen was of bullet-resisting glass, being made of four separate panels. Of these, three were wired for electric heating, but only the one immediately in front of the pilot was connected up. 10-mm. external armour was fitted over the fuselage skin forward of the windscreen. This extended across the fuselage and down each side to a level with the cockpit side windows. The pilot was protected from stern attacks by the usual shaped seat normally found in bombers. The engines were protected, as is usual in B.M.W. 801s in fighters, by two rings of armour round the nose.

Internal Equipment

Cockpit: The cockpit of this aircraft presents an entirely different appearance from that of previous Ju.88 fighters. This is due in part to the removal and the repositioning of the guns, also to the new instrument layout.

De-icing: De-icing of the main and tailplanes is by hot air. This is normally supplied through a muff fitted around the exhaust stubs, but in this aircraft it could not be traced. There was, however, provision for a petrol-fired heater (Kärcher Ofen) and a switch in the cockpit indicated that air could be supplied from this heater to either wings or fuselage.

In this case the heater was not installed but there was an air intake in the leading-edge of the port mainplane between the engine nacelle and fuselage, with an exit slightly further outboard on the wing upper surface. The piping has not yet been traced out, but presumably it will lead to the heater position.

The propeller de-icing is similar to that on the Me 410, a plain pipe delivering fluid direct onto the blade-root, whereas previously a pipe has led to a slinger-ring, from which a smaller pipe passed through the blade to emerge on the leading-edge about 18 inches from the root.

Control Column: *The right-hand horn of the control column incorporates fighter-type gun control buttons and also a small sliding switch which operates the elevator trimming tabs.*

Tankage: *The wing-tanks and the tank in the rear bomb-bay are normal. Due to the re-positioning of the forward firing armament, however, the space available in the forward bomb-bay only permits a tank of reduced size which is estimated to contain about 100 gallons. The total internal tankage of the aircraft is therefore restricted to approximately 620 gallons as compared with 790 gallons in the Junkers Ju.88 C-6 fighter.*

Undercarriage: *The frame is of welded cast-steel, as found in recently crashed Ju 188 bombers, instead of light alloy.*

Cable Cutter: *No balloon cable-cutter was fitted to this aircraft.*

Radar & Special Equipment: *The German A.I. (Airborne Interception) apparatus, the Lichtenstein FuG 220 (Model SN-2) is fitted.*

The aerial array consists of four di-poles with reflectors, and is mounted on right-angle arms projecting forward of the nose. So far, the operating frequency has been found to be of the order of 90 Mc/s.

A new installation for homing on to Allied Radar, (e.g. 'Monica") is also installed, comprising a receiver and a cathode-ray tube indicator, designated the FuG 227.

The azimuth aerials for the FuG 227 are mounted projecting forward, toed out, from the leading edge of the port and starboard mainplanes, at a point approximately three-quarters distant from the fuselage in each case. The elevation aerials are located above and below the starboard wing slightly outboard and behind the azimuth position.

The strikingly large diameter of the mounting rods and the aerial elements of both the FuG 220 and FuG 227 will serve as an excellent recognition feature. FuG 25A I.F.F. (Identification Friend-or-Foe) is fitted. There is provision for the Radio Altimeter, the FuG 101A, but the transmitter and receiver units are not installed in the special recessed panels in the underside of the port wing towards the tip.

Communication

Equipment: *The now standardised bomber installation, the FuG 10P, is fitted giving R/T (Short-wave) and W/T (Long-wave) with a PeGe 6 (Peilgerät) automatic Direction-Finder; the latter being coupled to the new form of combined D/F and*

This shot of Ju.88 TP190 clearly shows the 'antler' or Hirschgeweih aerials of the FuG 220 radar and the RAF applied 'P' prototype marking.

Inspection by the troops!

repeater compass ("S' Compass installation). Only the long-wave transmitter S.L. had its 'click-stop' frequency settings engaged, the frequencies being 377 and 468 Kc/s.

The pilot's V.H.F. R/T set proved to be a new type of designation, namely the FuG 16 ZY. The 'Z' component, which is the navigational attachment, is however missing from its mounting frame, but the loop aerial is mounted on the underside of the fuselage. A new semi-whip type aerial protrudes from the underside of the cockpit, being of tapered tubular construction terminating in a semi-stiff stranded portion, the overall length being 90-cm. This aerial is connected to a matching unit, type AAG 16 E3, which is a new number and is probably associated with the 'Y' characteristic of this FuG 16ZY installation.

The FuG 16 ZY receiver and transmitter were both 'click-stop' set at 40 Mc/s. It is interesting to note that a further pair of settings were engaged outside the calibrated limits of the frequency scale at positions which would be equivalent approximately to 42.475 and 42.525 Mc/s if the extended scale readings are assumed to be linear.

The blind-landing equipment, type FuBl 2F, is fitted, this being the electrically remote-controlled version now normally encountered in the Ju.88 aircraft.

Crew: Three. Prisoners of War. Uninjured. Only two of the crew were wearing a parachute and a single-seater dinghy.

Two days later, on 15 July Wing Commander 'Roly' Falk flew the aircraft – escorted by a pair of Spitfires – from Woodbridge to Farnborough via Hatfield. The aircraft was allocated serial TP190 and later Air Ministry number 231.

Study of this aircraft's radars made it possible, within ten days of its capture, for the RAFs Bomber Command to jam the Luftwaffe's SN-2 equipment, while the removal of Monica tail warning radar from British bombers rendered the German Flensburg equipment useless.

Soon after its capture this Ju.88 was repainted with German markings and the code 3K-MH and used in the making of an AI Mk. 10 radar training film. It was transferred to the Enemy Aircraft Flight, Central Fighter Establishment, Tangmere, on 17 May, 1945. This Ju.88 was further tested at the RAE by Flt Lt D. G. M. Gough on 11 October, and it was on exhibition at Farnborough in that and the following month before being transferred to No. 47 MU Sealand for storage.

Chapter Four

V-for-Vengence

Crossbow was the code name of the campaign of Anglo-American *'...operations against all phases of the German long-range weapons programme - against research and development of the weapons, their manufacture, transportation and their launching sites, and against missiles in flight'*. The original code name Bodyline was replaced with Crossbow on 15 November 1943.

Officials debated the extent of the German weapons' danger; some viewed the sites as mere decoys to divert Allied bombers, while others feared chemical or biological warheads.

At the request of the British War Cabinet, on 19 April 1944 Dwight Eisenhower directed that Crossbow attacks must have absolute priority over all other air operations, including *'...wearing down German industry and morale for the time being'*, which he confirmed after the V-1 assault began on the night of June 12/13, 1944: *'...with respect to Crossbow targets, these targets are to take first priority over everything except the urgent requirements of the Overlord battle; this priority to obtain until we can be certain that we have definitely gotten the upper hand of this particular business'*.

Carl Spaatz responded on June 28 to *'...complain that Crossbow was a 'diversion' from the main task of wearing down the Luftwaffe and bombing German industry'* for the Combined Bomber Offensive, and to recommend instead that Crossbow be a secondary priority since *'...days of bad weather over Germany's industrial targets would still allow enough weight of attack for the rocket sites and the lesser tactical crises'*. By 10 July Tedder had published a list of Crossbow targets which assigned thirty to RAF Bomber Command, six to Tedder's tactical forces, and sixty-eight to Spaatz's USSTAF; after which Spaatz again complained.

Vergeltungswaffen Eiener
On 22 August 1943 an object had crashed in a turnip field on the Island of Bornholm in the Baltic, roughly half-way between Germany and Sweden. It was a small pilotless aircraft bearing the number `V83`, and it was promptly photographed by the Danish Naval Officer-in-Charge on Bornholm, Lieutenant Commander Hasager Christiansen.

He also made a sketch, and noted that the warhead was a dummy made of concrete. He sent copies of his photographs and sketch to Commodore Paul Morch, Chief of the Danish Naval Intelligence Service, who forwarded them to British Intelligence.

At first, the British were not sure what had been found. From the sketch it was about 4 metres long, and it might have been a rather larger version of the HS.293 glider bomb that KG 100 was now using against warships in the Mediterranean.

This had been preceeded by a report originated on 12 August to Professor Reginald Victor Jones, CH CB CBE FRS, the Assistant Director of Intelligence (Science) from an officer in the Army Weapons Office who had told some weeks earlier of the plan for winged rockets.

In 1936 Jones took up the post at the RAE Farnborough. Here he worked on the problems associated with defending Britain from an air attack.

In September 1939, the British decided to assign a scientist to the Intelligence section of the Air Ministry. No scientist had previously worked for an intelligence service so this was unusual at the time. Jones was chosen and quickly rose to become Assistant Director of Intelligence (Science) there. During the course of the Second World War he was closely involved with the scientific assessment of enemy technology, and the development of offensive and counter-measures technology.

He was briefly based at Bletchley Park in September 1939, but returned to London (Broadway) in November, leaving behind a small specialised team.

Jones' first job was to study 'new German weapons', real or potential. The first of these was a radio navigation system which the Germans called Knickebein. This, as Jones soon determined, was a development of the Lorenz blind landing system and enabled an aircraft to fly along a chosen heading with useful accuracy.

At Jones' urging, Winston Churchill ordered up an RAF search aircraft on the night of 21 June 1940, and the aircraft found the Knickebein radio signals in the frequency range which Jones had predicted. With this knowledge, the British were able to build jammers whose effect was to 'bend' the Knickebein beams so that German bombers spent months scattering their bomb loads over the British countryside. Thus began the famous 'Battle of the Beams' which lasted throughout much of the war, with the Germans developing new radio navigation systems and the British developing countermeasures to them. Jones frequently had to battle against entrenched interests in the armed forces, but, in addition to enjoying Churchill's confidence, had strong support from, among others, Churchill's scientific advisor F D Lindemann and the Chief of the Air Staff, Sir Charles Portal.

Professor Reginald Victor Jones, CH CB CBE FRS,(b. 29 September 1911 – d.17 December 1997)

His report into the new pilotless aircraft - known as Phi7 - was much more specific. It said it was being tested at Peenemünde, but he knew nothing about it as it was not an Army project. In addition there was a rocket projectile known as A4.

A further report came in from a French agent via the famous 'Alliance' network headed by Marie-Madeleine Fourcade, who had forwarded it to Kenneth Cohen in London.

'*On the island of Usedon (north of Stettin) are concentrated laboratories and scientific research services to improve existing weapons and perfect new ones. The island is very closely guarded. Research is concentrated on:*
(a) bombs and shells guided independently of the laws of ballistics.
(b) a stratospheric shell.
(c) the use of bacteria as a weapon.
Kampfgruppe KG 100 is now experimenting with bombs guided from the aircraft by the bomb aimer. These bombs could be guided from such a distance that the plane could remain out of range of AA fire. Accuracy is perfect if the aircraft does not have to defend itself against fighters'.

Oberst Max Wachtel, commander of Flak Regiment 155(W), which was responsible for the V-1 offensive.

The information about KG 100 having used guided bombs which had been developed at Peenemünde against warships in the Mediterranean was correct: Bacterial warfare was an aspect which British Intelligence were also on the watch for, but what interested them most was the 'stratospheric shell.' This, and much of the internal evidence, with its 80 kilometres into the stratosphere, pointed to the rocket and had supplied a wealth of detail about the launching organisation in France. They learned for the first time of the existence of Colonel Max Wachtel and his Lehr-und-Erprobungskommando that was now to be formed into Flak Regiment I55W, and his technical adviser Major Sommerfeld.

Some of the figures in the report were frightening: 108 catapults

The Peenemünde Army Research Centre - *Heeresversuchsanstalt Peenemünde (HVP)* as seen here in an Allied Reconnaissance photograph was founded in 1937 as one of five military proving grounds under the German Army Weapons Office *(Heeres Waffenamt)* It is widely regarded as the birthplace of modern rocketry and spaceflight. *(USAAF)*

rising to more than 400, with 50 to 100 bombs sufficient to destroy London, and with a range up to 500 kilometres. What were they to make of it all?

Further clues came from an intercepted Enigma message sent on 7 September. The name 'Enigma' was given to several electro-mechanical rotor cipher machines used by the Germans for enciphering and deciphering secret messages. Of this particular message, the most significant part read:

'Luftflotte 3 again requests the immediate bringing up of Flak forces to protect ground organization Flak Zielgerät 76. The urgency of this is emphasized by the following facts: According to report of C. in C. West, Abwehr Station France reports the capture of an enemy agent who had the task of establishing at all costs the position of new German rocket weapon. The English, it is stated, have information that the weapon is to be employed in the near future and they intend to attack the positions before this occurs'.

The British Scientific Intelligence community immediately increased their interest in Flak Zielgerät 76 – shortened to FZG 76 - and it was not long before reports started to come in from Northern France about emplacements being built ten to twenty miles inland that were said to be for long range rockets or guns. The codename 'Flakzielgerät 76'— 'Flak aiming apparatus' helped to hide the nature of the device. Initially, British experts were sceptical of the V-1 because they had considered only solid fuel rockets, which could not attain the stated range of 1,000 kg over 209 km. However they later considered other types of engine, and by the time German scientists had achieved the needed accuracy to deploy the device as a weapon, British intelligence had a very accurate assessment of it.

On 25 October 1943, in a meeting with Prime Minister Churchill, Professor Jones concluded: *'Much information has been collected. Allowing for the inaccuracies which often occur in individual accounts, they form a coherent picture which, despite the bewildering effect of propaganda, has but one explanation: the Germans have been conducting an extensive research into long range rockets at Peenemiinde. Their experiments have naturally encountered difficulties which may still be holding up production, although Hitler would press the rockets into service at the earliest possible moment; that moment is probably still some months ahead. It would be unfortunate if because our sources had given us a longer warning than was at first appreciated, we should at this stage discredit their account.*

There are obvious technical objections which, based on our own experience, can be raised against the prospect of successful rockets, but it is not without precedent for the Germans to have succeeded while we doubted: the beams are a sufficient example.

It is probable that the German Air Force has been developing a pilotless aircraft for long range bombardment in competition with the rocket, and it is very possible that the aircraft will arrive first.

On 28 October reports came in from the French Resistance of new German constructions near the Channel coast detailing work proceeding at Bois Carré near Yvrench, and that a rough plan made by a workman showed *'...a concrete platform with centre axis pointing directly to London'*. It described the construction of various buildings on the site, including one which contained no metal parts; subsequent reports modified this to no magnetic parts. The information was confirmed by photographic sortie E/463 of 3 November, which showed that the most prominent features were ski-shaped buildings 240-270 feet long, from which the sites were promptly named. If the non-magnetic hut story were true, this would indicate some kind of magnetic directional control in the missile. The

Original caption: *'Robot ski site at Yvrench, France photographed on 9 November 1943 shows ski in process of construction. Complete sites call for three skis, which are long low buildings of heavy concrete, used for storage. Reason for curved and not yet known'* In fact, these were flying bomb or as they were colloquially known, 'doodlebug' locations. (USAAF).

German troops manhandle a V-1 on it's ground trolley before launch.

non-magnetic hut had a low, wide arch, which suggested that something with wings had to be wheeled through it.

Eventually the weather improved over Zempin and Peenemünde so on 28 November 1943 a de Havilland Mosquito photo-reconnaissance aircraft, flown by Squadron Leader John Merifield was able to take pictures. They showed the same building at Zempin that had been observed at the ski-sites, giving final proof that they were intended for the FZG.76. They also showed the catapult, which was also identified at Peenemünde - previously misinterpreted as 'sludge-pumps'. The icing on the cake was that just as the Mosquito flew over the Germans had a flying bomb in position on one of the catapults.

An RAF and USAAF bombing campaign was immediately initiated against the 96 ski sites located - such was the effort against them that by the end of April 1944 the Chiefs of Staff considered that the threat would be neutralised by the end of the month.

However, the Germans then moved the goalposts - aerial photographs showed that they were now building what was termed 'modified ski sites' from which the now familiar skis were missing and which could be much more rapidly erected. They also developed mobile launchers.

The first ten bombs were launched on the night of 12/13 June. Five crashed shortly after launching, and a sixth one went missing; of the remaining four, one fell in Sussex, and the others near Gravesend, near Sevenoaks and at Bethnal Green. This was just a prelude of what was to come.

As it was the Germans quickly overcame their deficiencies, and on 15 June the campaign opened in earnest, and the people of London tended to feel that the Government had been taken unawares.

In the 24 hours beginning at 22.30 on 15 June, Wachtel launched more than 200 flying bombs, of which 144 crossed our coasts and 73 reached Greater London. 33 bombs were brought down by the defences but eleven of these came down in the built-up area of Greater London.

Within a few days of the opening of the bombardment, a new problem arose. As with any remote device, the German's needed to know where the missiles were landing so as to be able to adjust and correct settings in the guidance system. Therefore, German intelligence was requested to obtain this impact data from their agents in Britain. However, most German agents in Britain had been turned, and were acting as double agents under British control.

On 16 June 1944, British double agent Garbo (Juan Pujol) was requested by his German controllers to give information on the sites and times of V-1 impacts, with similar requests made to the other German agents in Britain, Brutus (Roman Czerniawski) and Tate (Wulf Schmidt). If given this data, the Germans

A remarkable photograph showing a Fi.103 flying bomb falling on the Piccadilly area of London in July 1944. *(USAAF)*

A German Luftwaffe Heinkel He.111 H-22. This version could carry FZG 76 (V-1) flying bombs, but only a few aircraft were produced in 1944.

would be able to adjust their aim and correct any shortfall. However, there was no plausible reason why the double agents could not supply accurate data; the impacts would be common knowledge amongst Londoners and very likely reported in the press, which the Germans had ready access to through the neutral nations.

While the British decided how to react, Pujol played for time. On 18 June it was decided that the double agents would report the damage caused by V-1s fairly accurately and minimise the effect they had on civilian morale. It was also decided that Pujol should avoid giving the times of impacts, and should mostly report on those which occurred in the north west of London, to give the impression to the Germans that they were overshooting the target area.

While Pujol downplayed the extent of V-1 damage, trouble came from Ostro, an Abwehr agent in Lisbon who pretended to have agents reporting from London. He told the Germans that London had been devastated and had been mostly evacuated due to enormous casualties. The Germans could not perform aerial reconnaissance of London, and believed his damage reports in preference to Pujol's. They thought that the Allies would make every effort to destroy the V-1 launch sites in France. They also accepted Ostro's impact reports. Due to Ultra however, the Allies read his messages and adjusted for them.

Following the Liberation of France, a number of V-1 assembly sites were overran by the Allies. Here troops inspect a partially assembled example.

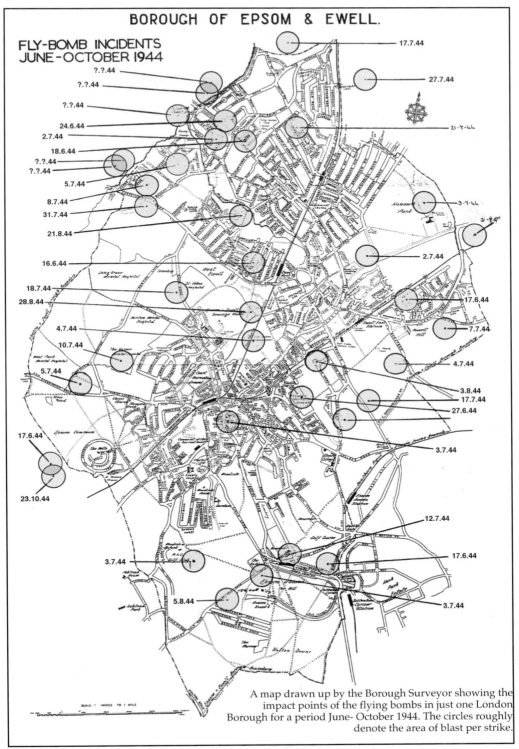

A map drawn up by the Borough Surveyor showing the impact points of the flying bombs in just one London Borough for a period June- October 1944. The circles roughly denote the area of blast per strike.

A certain number of the V-1s fired had been fitted with radio transmitters, which had clearly demonstrated a tendency for the V-1 to fall short. Oberst Max Wachtel compared the data gathered by the transmitters with the reports obtained through the double agents. He concluded, when faced with the discrepancy between the two sets of data, that there must be a fault with the radio transmitters, as he had been assured that the agents were completely reliable. It was later calculated that if Wachtel had disregarded the agents' reports and relied on the radio data, he would have made the correct adjustments to the V-1's guidance, and casualties might have increased by 50% or more.

Each conventional launch site could theoretically launch about fifteen V-1s per day, but this rate was difficult to achieve on a consistent basis; the maximum rate achieved was eighteen. Overall, only about 25% of the V-1s hit their targets, most being lost because of a combination of defensive measures, mechanical unreliability, or guidance errors. With the capture or destruction of the launch facilities used to attack England, the V-1s were employed in attacks against strategic points in Belgium, primarily the port of Antwerp.

Launches against Britain were met by a variety of countermeasures, including barrage balloons, Hawker Tempest and even Gloster Meteor jet fighters. These measures were so successful that by August 1944 about 80% of V-1s were being destroyed. The Meteors, though fast enough to catch the V-1s, suffered frequent cannon failures, and accounted for only thirteen. In all, about 1,000 V-1s were destroyed by aircraft.

The intended operational altitude was originally set at 2,750 m (9,000 ft). However, repeated failures of a barometric fuel-pressure regulator led to it being changed in May 1944, which had the effect of halving the operational height, thus bringing V-1s into range of the Bofors guns used by Allied AA units.

While the conflict was ongoing, the Allies did everything they could to destroy the missiles - but the moment that territory was captured where these weapons were located, the same amount of effort was spent in learning all they could about them. Given the unmanned nature of these weapons it was not surprising that the Allies were interested in the nuts and bolts of the machines. No sooner than parts of Europe had been liberated, teams of scientists and technicians investigated whatever was found. These were the first practical cruise missiles and were capable of incredible development.

This was clear when it was discovered that the V-1 were also air-launched. Most V-1s were launched from static sites on land, but from July 1944 to January 1945 the Luftwaffe launched many from modified Heinkel He.111 H-22s of the Luftwaffe's Kampfgeschwader 3 (3rd Bomber Wing, the so-called 'Blitz Wing') flying over the North Sea. Apart from the obvious motive of permitting the bombardment campaign to continue after static ground sites on the French coast were lost, air-launching gave the Luftwaffe the opportunity to outflank the increasingly effective ground and air defences put up by the British against the missile. To minimise the associated risks (primarily radar detection), the aircrews developed a tactic called 'lo-hi-lo': the He.111s would, upon leaving their airbases and crossing the coast, descend to an exceptionally low altitude. When the launch point was neared, the bombers would swiftly ascend, fire their V-1s, and then rapidly descend again to the previous 'wave-top' level for the return flight. Research after the war estimated a 40% failure rate of air-launched V-1s, and the He.111s used in this role were extremely vulnerable to night fighter attack, as the launch lit up the area around the aircraft for several seconds. The combat potential of air-launched V-1s dwindled at about the same rate as that of the ground-launched missiles, as the British learned more of the weapon and developed increasingly effective defence tactics. For example, during Operation 'Martha', KG 3's He.111H-22s fired 45 V-1s at Britain in a single concerted strike on Christmas Eve 1944, with just one missile getting through to hit a target.

German test pilot and aviatrix Hanna Reitsch (b. 29 March 1912 – d. 24 August 1979)

A captured *Reichenberg* - one of the V1s fitted with flight controls - is inspected by British Army personnel.

Late in the war, several air-launched piloted V-1s, known as *Reichenbergs*, were built, but never used in combat. Hanna Reitsch made some flights in the modified V-1 *Reichenberg* when she was asked to find out why test pilots were unable to land it and had died as a result. She discovered, after simulated landing attempts at high altitude where there was air space to recover, that the craft had an extremely high stall speed and the previous pilots with little high speed experience had attempted their approaches much too slowly. Her recommendation of much higher landing speeds was then introduced in training new Reichenberg volunteer pilots.

There were plans, not put into practice, to use the Arado Ar 234 jet bomber to launch V-1s either by towing them aloft or by launching them from a 'piggy back' position atop the aircraft. In the latter configuration, a pilot-controlled, hydraulically operated dorsal trapeze mechanism would elevate the missile on the trapeze's launch cradle some eight feet clear of the 234's upper fuselage. This was necessary to avoid damaging the mother craft's fuselage and tail surfaces when the pulse jet ignited, as well as to ensure a 'clean' airflow for the Argus motor's intake.

A somewhat less ambitious project undertaken was the adaptation of the missile as a 'flying fuel tank' (Deichselschlepp) for the Messerschmitt Me.262 jet fighter, which was initially test-towed behind an He.177A Greif bomber. The pulse-jet, internal systems and warhead of the missile were removed, leaving only the wings and basic fuselage, now containing a single large fuel tank. A small cylindrical module, similar in shape to a finless dart, was placed atop the vertical stabiliser at the rear of the tank, acting as a centre of gravity balance and attachment point for a variety of equipment sets. A rigid tow-bar with a pitch pivot at the forward end connected the flying tank to the Me.262. The operational procedure for this unusual configuration saw the tank resting on a wheeled trolley for take-off. The trolley was dropped once the combination was airborne, and explosive bolts separated the towbar from the fighter upon

One device that interested the Allies was the *'Diechselschepp'* airborne trailer that was flight tested as a means on increasing the bombload of the Me.262. The device used a four metre long cylindrical towbar that was fitted to the rear of the aircraft by a swivel attachment. At the other end of the towbar, on a two-wheeled trolley was a bomb fitted with wings from a Fi.103 V-1. After take-off the wheels were jettisoned. Once over the target the pilot put the 262 into a shallow dive and, using the Revi gunsight he released the bomb, at the same time disengaging the towbar and wing. A number of trials were undertaken, but after one when the bomb broke away and on another the pilot was forced to land with the bomb still attached, the Germans realised that the project was unsafe, and it was abandoned.

This was not the only use of an 'aerial trailer' - a similar method was tried to increase the radius of action of the Ar.234 bomber by towing a V-1 flying bomb with the warhead, engine and tailplane removed and a wheeled undercarriage fitted, to carry extra fuel. Hitler took an interest in the tests, as Reichsminister Albert Speer noted in his diary on 5 November 1944: *"Reported to Hitler on the experiments intended with the 234: to employ it as a valuable super-fast warplane, but connecting it to an engineless V-1 weapon and using the fuel container of the V-1 as an additional tank for the outward journey, to give the 234 the widest possible range and to enable it to carry a considerably heavier external bomb load. Hitler expects these experiments to be concluded with all possible speed and tested."* The idea was not a success, probably owing to stability problems, and the scheme was never tried in action.

exhaustion of the tank's fuel supply. A number of test flights were conducted in 1944 with this set-up, but inflight 'porpoising' of the tank, with the instability transferred to the fighter, meant the system was too unreliable to be used. An identical utilisation of the V-1 flying tank for the Ar 234 bomber was also investigated, with the same conclusions reached. Some of the 'flying fuel tanks' used in trials utilised a cumbersome fixed and spatted undercarriage arrangement, which (along with being pointless) merely increased the drag and stability problems already inherent in the design.

One variant of the basic Fi.103 design did see operational use. The progressive loss of French launch sites as 1944 proceeded and the area of territory under German control shrinking meant that soon the V-1 would lack the range to hit targets in England. Air-launching was one alternative utilised, but the most obvious solution was to extend the missile's range. Thus the F-1 version developed. The weapon's fuel tank was increased in size, with a corresponding reduction in the capacity of the warhead. Additionally, the nose-cones and wings of the F-1 models were made of wood, affording a considerable weight saving. With these modifications, the V-1 could be fired at London and nearby urban centres from prospective ground sites in the Netherlands. Frantic efforts were made to construct a sufficient number of F-1s in order to allow a large-scale bombardment campaign to coincide with the Ardennes Offensive, but numerous factors, including the bombing of the factories producing the missiles, shortages of steel and rail transport, and the chaotic tactical situation Germany was facing delayed the delivery of these long-range V-1s until February/March 1945. Beginning on 2 March 1945, a little more than three weeks before the V-1 campaign ended for good, several hundred F-1s were launched at Britain from Dutch sites under Operation 'Zeppelin'.

Vergeltungswaffen Zwei

In recent years the Vergeltungswaffe Zwei (Vengeance Weapon 2) has been portrayed by many as Germany's last desperate hope. This is far from the truth. Planning and production of the rocket weapon was extensive and wide ranging. It was the culmination of years of strategic thinking.

The V-2 was an unmanned, guided, ballistic missile. It was guided by an advanced gyroscopic system that sent signals to aerodynamic steering tabs on the fins and vanes in the exhaust. It was propelled by an alcohol mixture of 75% ethyl alcohol and 25% water, and liquid oxygen fuel. The two liquids were delivered to the thrust chamber by two rotary pumps, driven by a steam turbine. The steam turbine operated at 5,000 rpm on two auxiliary fuels, namely hydrogen peroxide (80%) and a mixture of 66% sodium permanganate with water 33%. This system generated about 55,000 lbs of thrust at the start, which increased to 160,000 lbs when the maximum speed was reached. The motor typically burned for 60 seconds, pushing the rocket to around 4,400 ft/second. It rose to an altitude of 52 to 60 miles and had a range of 200 to 225 miles. The V-2 carried an explosive warhead (amatol Fp60/40) weighing approximately 1 ton that was capable of flattening a city block. It was first fired operationally on 8 September 1944 against Paris then

London, this was the beginning of the V-2 campaign.

One name is inexorably linked to the V-2. Wernher Magnus Maximilian von Braun was born on 23 March 1912, in Wirsitz, a town in the eastern German province of Posen. From childhood, Wernher revealed an interest in both science and music. At age 11 he enrolled in the Französisches Gymnasium that had been established two centuries earlier by Fredrick the Great. There, the boy showed only a modest ability in mathematics and physics, subjects in which he would later excel. In 1928 Wernher's father placed him in the progressive Hermann Lietz schools. Wernher's grades and abilities improved. Oberth's book captured the young boy's attention. However, von Braun soon learned that he would have to excel in mathematics to even understand the concepts and principles in the book. Even during these younger years of his life, von Braun was experimenting with rockets and propulsion.

A pivotal point occurred for the then 18-year-old von Braun when he entered the Technische Hochschule in the Berlin district of Charlottenburg. While in Berlin, von Braun's interest in astronomy and space travel continued to grow. He had become acquainted with Hermann Oberth, writer and spaceflight promoter Willy Ley, and rocket experimenters Rudolf Nebel and Johannes Winkler. He also followed the solid-fuel exploits of Max Valier. Von Braun quickly joined the *Verein für Raumschiffahrt* (VfR) and was soon participating in rocket experiments in Berlin.

The development of the rocket had its roots in the enthusiastic amateur German rocket societies, which cultivated emerging specialists such as Wernher von Braun; however, it was the German military, using the emerging technology as a weapon for war, which shaped the V-2. The Ballistics and Munitions Branch was solely interested in collecting real scientific data on rocket propulsion. The problem was usually that these groups drew a huge amount of publicity, something the Army wanted to avoid at all costs.

The Peenemünde complex began work on the military weapon, the A-4, with von Braun in charge of technical development. After several tries, an A-4 missile was successfully launched on 3 October 1942. Much was still unfinished though - a single successful launch did not translate into a proven weapon system. It would be two years later that the first A-4/V-2s were operationally deployed.

During these fast-moving early development stages of Peenemünde's growth, the multitude of German scientists and engineers were well served by the young and resourceful von Braun. His ability to put the right personnel in key positions, the ability to streamline research efforts, head off disputes, secure materials and von Braun's own exuberance for the A-4 project were key in Peenemünde's success.

The first real inkling the Allies had of this new wonder weapon occured in June 1943, when following a photographic reconnaissance sortie over Peenemünde an object throught to be a rocket was spotted by Professor R V Jones. A further sortie over the Research Establishment occurred on 26 June was flown by Flt Sgt E P H Peek - the flight obtained pictures that clearly showed a rocket and an arrangement of 'Giant Wurtzburg' radar antennas.

Given that Peenemünde was also associated with the V-1 programme, it was agreed that the facility should be attacked by Bomber Command on the heaviest possible scale on the first occasion when conditions were suitable. The attack took place on 17/18 August 1943.

The effects of the raid in delaying the rocket programme have been variously estimated from four weeks to six months. The RAF killed some key German personnel, such as Dr. Walter Thiel, who was responsible for rocket jet design, and burnt up many of the production drawings. They also did enough damage to the station itself to make the Germans decide that they ought to remove both development and production facilities to other places. Taken altogether, it was

From left to right, General Dr Walter Dornberger, General Friedrich Olbricht (with Knight's Cross), Major Heinz Brandt, and Wernher von Braun (in civilian suit) at Peenemünde, in March 1941. Von Braun joined the Nazi Party on 12 November 1937 and was issued with membership number 5,738,692. He also joined the SS with membership number 185,068. He began as an *Untersturmführer* (Second Lieutenant) and was promoted three times by Reichsfuehrer SS Heinrich Himmler, the last time in June 1943 to *SS-Sturmbannführer* (Major).

thought that the raid must have slowed the project by at least two months, and these would have been very significant because the rocket could then have been used almost simultaneously with the flying bomb in 1944, and from a shorter range since the Germans still held northern France and Belgium. The UK would have had to face two threats at the same time.

Ever since the attack of 17/18 August 1943 on Peenemünde, the long-range rocket seemed to have been eclipsed by the flying bomb. Through the use of Ultra, the British had a ringside seat at the Baltic trials of the V-weapons.

Ultra was the designation adopted by British military intelligence in June 1941 for wartime signals intelligence obtained by breaking high-level encrypted enemy radio and teleprinter communications at the Government Code and Cypher School (GC&CS) at Bletchley Park. Ultra eventually became the standard designation among the western Allies for all such intelligence. The name arose because the intelligence thus obtained was considered more important than that designated by the highest British security classification then used (Most Secret) and so was regarded as being Ultra secret. Several other cryptonyms had been used for such intelligence.

Much of the German cipher traffic was encrypted on the Enigma machine. Used properly, the German military Enigma would have been virtually unbreakable; in practice, shortcomings in operation allowed it to be broken. The term 'Ultra' has often been used almost synonymously with 'Enigma decrypts'. However, Ultra also encompassed decrypts of the German Lorenz SZ 40/42 machines that were used by the German High Command, and the Hagelin machine and other Italian and Japanese ciphers and codes such as PURPLE and JN-25.

The RAF raid did not drive all the experimental work from Peenemünde as had been expected; but some of it moved eastwards along the Baltic coast to Brüster Ort, and what appeared to be an out-station of Peenemünde was set up at an SS camp known as 'Heidelager` at Blizna near Debice, some 170 miles south of Warsaw. This was first located via some Ultra traffic, and it seemed to be primarily concerned with flying bomb trials over land.

Since British Intelligence already knew so much about the flying bomb,

An A-5 test rocket suspended beneath a He.111 for drop tests. The A-5 was a test-bed rocket for all principle features of the A-4. Equipped with the new Siemens control equipment, it was designed to execute commanded guidance during a ballistic trajectory and have the ability to do so in stable flight. The rocket would manoeuver into a ballistic attitude when the gyroscopes tilted in the desired direction of flight, which would cause the autopilot to send signals to the servos attached to the exhaust vanes.

studies of the activity in Poland could add little to their knowledge in this respect; but there were features about the messages that could not be explained by assuming that the flying bomb alone was involved. In March 1944 the Polish Intelligence Service reported that trials were being conducted at Blizna of a missile with a range of 10 kilometres which made a large crater, and that railway tank cars thought to contain liquid air were entering the establishment. Finally, there was one item from the Ultra traffic which could not be accounted for either by the flying bomb or a short-range missile, because one of the staff at Blizna was interested in a crater near Sidlice, some 160 miles away to the north-east, which was beyond the range of the flying bomb. This made British Intelligence think that they were once again on the trail of the rocket.

A photographic reconnaissance sortie of the Blizna area was flown on 5 May, 1944 when a flying bomb ramp was immediately recognized, but there was none of the other installations previously seen at Peenemünde thought to be associated with the rocket. Since the photographic cover was far from complete, it seemed that there must be another compound in which the rocket activity was taking place.

Then, on the same day that the first flying bomb fell in England, a large rocket from Peenemünde fell in Sweden. The British scientific community later learned that this was an experimental hybrid consisting of a genuine A4 rocket fitted with an elaborate radio guidance system primarily designed for a smaller rocket known as 'Wasserfall' which was being developed for anti-aircraft purposes. It took a few days before news of the incident reached London; the report from the British Air Attaché in Stockholm was accompanied by some rather badly focused photographs of the remains of the crashed missile taken by the Swedish General Staff. These showed components that were not associated with the flying bomb, and there was speculation that another sample of the Germans' latest V-weapon had gone astray. Two Air Technical Intelligence Officers - Squadron Leaders Burder and Wilkinson – flew to Sweden with a request to the Swedish General Staff that they might be allowed to inspect the debris. Their first report back to Air Ministry expressed surprise at the amount of electronic control equipment incorporated in the rocket, and gave their opinion that such expensive complexity would only be justified if the warhead were to weigh at least ten thousand pounds.

Further confirmation of a large warhead seemed to come from the aerial photograph of Blizna, which showed a large crater a few kilometres from the S.S. camp. Much additional evidence was now being provided by the Poles: they had set up an organisation for beating the Germans to the scenes of crashed missiles, and they had analysed the liquid recovered from one incident, and found it to be highly concentrated hydrogen peroxide.

By 27 June they had reported that the missile was about 40 feet long and 6 feet in diameter, agreeing with the dimensions that had been measured on the Peenemünde photographs.

They also found pieces of radio equipment, including a transmitter on a wavelength of about 7 metres and a receiver for about 14 metres; these were investigated by Professor Janusz Groszkowski, later the President of the Polish Academy of Science. Shortly afterwards the Poles reported on the construction of the main jet, and, from the dimensions they gave, this was identical with the rocket unit that had fallen in Sweden. During July the Poles also offered to send the pieces they had collected to the UK if an aircraft could pick it up, and this operation was put in hand.

New evidence came to light as Allied Forces advanced in Normandy when traces of an organization for storing and launching rockets from south of Caen was discovered. One of the sites was found near the Chateau du Molay, west of Bayeux - this was simply a tree-flanked stretch of road into which concrete platforms had been set, and on either side of which parallel loop roads had been made among the trees. These were very similar to a pattern seen some months before on the foreshore at Peenemünde, which had hitherto made no sense. Here now was an explanation: the pattern of the roads at the Chateau du Molay had been laid out on the sands at Peenemünde to see whether the proposed curves in the loop roads could be negotiated by whatever transporters were to carry the rockets.

A reconnaissance photo of V-2 launch site at Peenemünde. The rocket stood in the middle of the oval earth bank.

After many trials and tribulations the rocket parts began to arrive at Farnborough in the UK, where reconstruction of the pieces started and where eventually, after much discussion, it was decided that the rocket had a warhead weighing one ton.

As early as August 1944 it was known that the Germans were planning on firing 800 rockets a month. An obvious countermeasure to the rocket was to bomb the factories that were making it. The British had found the three main factories, Peenemünde, the Zeppelin Works at Friedrichshafen, and the Rax Werkes at Wiener Neustadt, each of which was to produce 300 rockets per month, and bombed all three. This caused the Germans to concentrate the production of V-1 and V-2s in a great underground factory at Nordhausen, which from August 1944 onwards produced 600 rockets a month. At the time all the British knew was that there was an underground factory the 'M(ittel) Works' somewhere in central Germany (actually it was in the Harz mountains north of Nordhausen) operating in conjunction with a concentration camp 'Dora'.

In actuality, the concentration camp Dora-Mittelbau was established in 1943

A test V-2 is launched.

in the Harz mountain near Nordhausen for the sole purpose of producing the V-l and V-2. The tunnels under the Harz Mountains had been used by the *Wirtschaftliche Forschungsgesellschaft, GMbH* known as WiFo – the Economic Research Company - which was charged with the construction and operation of solid fuel - both natural and synthetic - storage depots.

WiFo had been created in 1934 to obtain raw material which it stored in bomb-proof underground stock-rooms and was helped by IG Farben to build the tunnels in the Kohnstein. In 1943 there were two big tunnels; WiFo planned two further parallel tunnels, each 1.8 kilometers in length and a cross-section of 60-90 square metres. These would be joined by fifty other tunnels of 200 metres length. The work had already started in 1936. When the prisoners arrived the big B1 tunnel was finished, but not the A. Of the 50 interconnecting tunnels 42 were finished.

With the threats to missile production increasing, the Germans searched for new sites and found the underground stock-room of the Wifo in the Harz in July 1943. Soon 107 prisoners arrived from Buchenwald at the new Dora Camp. On 2 September a further 1,212 prisoners arrived from Buchenwald, and on 1 January 1944 the population of Dora was well over 10,000.

Hitler decided to empty the tunnels despite the protestations of WiFo. Production of the V-2 was the responsibility of Mittelbau, who reported directly to Albert Speer's Ministry of Armament. Everything was carried out in total secrecy and was one reason for using prison workers.

The first prisoners had the hard task of completing the remaining eight interconnecting tunnels. Mortality was high. They also had to complete the air conditioning, water and electrical installations. Initially they were billeted in the tunnels, and it was only in the spring of 1944 that barracks were built.

From the spring of 1944 other programmes were added to the building of the V-2. Tunnels 0 to 20 were transferred to Junkers for the production of jet engines. Later tunnels 43 to 46 were transferred to the production of winged bombs.

From the spring of 1944, due to further Allied bomber attacks on the aircraft industry in February of that year, big new constructions started to make additional tunnels. Countless thousands of prisoners from the concentration camps were sent there. The object was to make about twenty big new tunnels in the Himmelberg not far from Kohnstein. The B3 Anhydrit tunnel was mainly built by Belgians, who were also employed in large numbers in building B12. This gigantic tunnel system was built to obtain 600,000 square metres of underground aircraft factory. Another project, the B1l, was proposed to realise a further 250,000 square metres of tunnel space north of Dora.

In October 1944 Dora became a completely independent concentration camp, with many external working groups: Ellrich, Harzungen, etc. In December 1944 the number of prisoners reached 30,000, including 14,000 in the Dora Werkes. However, little of the fine detail of this was known to the Allies at the time.

Nevertheless, reviewing all the evidence by the British Intelligence community created an amount of reasonable extrapolation that a two-stage rocket of about 150 tons starting weight could deliver a one ton warhead to nearly 3,000 miles range, with a probable error of ten miles in range and three miles in line. As Professor R V Jones went on record: *'This might be a feasible weapon for delivering a uranium bomb, should such a bomb become practicable. It would be almost hopeless to counter by attacks on the ground organisation, because the increased range would allow an almost unlimited choice of firing site, while the trajectory could be so varied that the firing point could not be deduced with sufficient accuracy for countermeasures.*

Production would probably take place underground. At the moment such a rocket could not be intercepted, but by the time it becomes a serious possibility it may itself be a target for smaller defence rockets fitted with predictors and homing devices: but these would depend upon adequate warning, and the defences might also be saturated by a salvo of long range rockets.'

Then the rockets started falling. Beginning in September 1944, over three thousand V-2s were launched by the Wehrmacht against Allied targets during the war, firstly London and later Antwerp and Liège. According to a number of sources, the attacks resulted in the deaths of an estimated 9,000 civilians and military personnel, while 12,000 forced laborers and concentration camp prisoners were killed producing the weapons

As Germany collapsed, teams from the Allied forces swiftly moved to capture key German manufacturing sites and items of equipment. One such mission was Operation Backfire, organised by the British authorities immediately after the end of hostilities in Europe, which was designed to completely evaluate the entire V-2 assembly, interrogate German personnel specialised in all phases of it, and then actually launch several missiles across the North Sea.

Vergeltungswaffen Drei
Before any of the ski sites had been discovered in 1943, British Intelligence found some large concrete constructions which from their deployment seemed likely to be used for launching missiles against London. Three of them, Lottinghem (behind Boulogne), Siracourt (near Saint Pol) and Equeurdreville (near Cherbourg), proved to be artificial caves for the assembly and launching of flying bombs.

Another two, Wizernes (also behind Boulogne) and Sottevast (behind Cherbourg), were to serve a similar function for the rockets, and one at Watten (behind Calais) was to be both a rocket launching point and a sheltered liquid oxygen plant.

The Allies located another great concrete structure at Mimoyecques, near Calais. The site was heavily bombed, and could probably never have come into action, but the Intelligence services did not deduce its exact purpose until the Allied forces overran it.

This was a supergun named 'V-3' working on the multi-charge principle whereby secondary propellant charges are fired to add velocity to a projectile. The V-3 was also known as the *Hochdruckpumpe* ('High Pressure Pump', HDP for short), which was a code name intended to hide the real purpose of the project. It was also known as Fleißiges Lieschen ('Busy Lizzie').

The weapon was planned to be used to bombard London from two large bunkers. The gun used multiple propellant charges placed along the barrel's length and timed to fire as soon as the projectile passed them, to provide an additional boost. Because of their greater suitability and ease of use, solid-fuel rocket boosters were used instead of explosive charges. These were arranged in symmetrical pairs along the length of the barrel, angled to project their thrust against the base of the projectile as it passed. This layout spawned the German codename *Tausendfüßler* ('millipede'). Unlike conventional rifled weapons of the day, the smoothbore gun fired a fin-stabilized shell, dependent upon aerodynamic rather than gyroscopic forces to prevent tumbling, which resulted in a lower drag coefficient.

In May 1943 Albert Speer, the Reich Minister of Armaments and War Production, informed Adolf Hitler of work that was being carried out to produce a supergun capable of firing hundreds of shells an hour over long distances. Long-range guns were not a new development, but the high pressure detonations used to fire shells from previous such weapons, including the Paris gun, rapidly wore out their barrels. In 1942, August Coenders, inspired by previous designs of multi-chamber guns, suggested that the gradual acceleration of the shell by a series of small charges spread over the length of the barrel might be the solution to the problem of designing very long-range guns. Coenders proposed the use of electrically activated charges to eliminate the problem of the premature ignition of the subsidiary charges experienced by previous multi-chamber guns. The HDP would have a smooth barrel over 100 metres long, along which a 97-kilogram finned shell (known as the Sprenggranate 4481) would be accelerated by numerous small low-pressure detonations from charges in branches off the barrel, each fired electrically in sequence. Each barrel would be 15 centimetres in diameter.

The gun was still in its prototype stages, but Hitler was an enthusiastic supporter of the idea and ordered that maximum support be given to its development and deployment. In August 1943 he approved the construction of a battery of HDP guns in France to supplement the planned V-1 and V-2 missile campaigns against London and the south-east of England.

The weapon needed barrels 127 metres long, so it could not be moved; it would have to be deployed from a fixed site. A study carried out in early 1943 had shown that the optimal location for its deployment would be within a hill with a rock core into which inclined drifts could be tunnelled to support the barrels.

The site was identified by a fortification expert, Major Bock of the Festungs-Pionier-Stab 27 of the Fifteenth Army LVII Corps based in the Dieppe area. A

The prototype V-3 cannon at Laatzig, Germany (now Poland) in 1942.

limestone hill near the hamlet of Mimoyecques, 158 metres high and 165 kilometres from London, was chosen to house the gun. It had been selected with care; the hill in which the facility was built is primarily chalk with very little topsoil cover, and the chalk layer extends several hundred metres below the surface, providing a deep but easily tunnelled rock layer. The chalk was easy to excavate and strong enough to provide tunnels without using timber supports. Although the site's road links were poor, it was only a few kilometres west of the main railway line between Calais and Boulogne-sur-Mer. The area was already heavily militarised; as well as the fortifications of the Atlantic Wall on the cliffs of Cap Gris Nez to the north-west, there was a firing base for at least one conventional Krupp K5 railway gun about five kilometres to the south in the quarries of Hidrequent-Rinxent.

Construction began in September 1943 with the building of railway lines to support the work, and excavation of the gun shafts began in October. The initial layout comprised two parallel complexes approximately 1,000 metres apart, each with five 'drifts' - or tubes - which were to hold a stacked cluster of five HDP gun tubes, for a total of 25 guns. The smoothbore design of the HDP would enable a much higher rate of fire than was possible with conventional guns. The entire battery would be able to fire up to 10 shots a minute, capable in theory of hitting London with 600 projectiles every hour. Both facilities were to be served by an underground railway tunnel of standard gauge, connected to the Calais-Boulogne main line, and underground ammunition storage galleries which were tuneled at a depth of about 33 metres. The western site was abandoned at an early stage after being disrupted by Allied bombing, and only the eastern complex was built.

The drifts were angled down at 50 degrees, reaching a depth of 105 m. Owing to technical problems with the gun prototype, the scope of the project was reduced; drifts I and II were abandoned at an early date and only III, IV and V were taken forward. They came to the surface at a concrete slab or Platte 30 metres wide and 5.5 metres thick, in which there were narrow openings to allow the projectiles to pass through. The openings in the slab were protected by large steel plates, and the railway tunnel entrances were further protected by armoured steel doors. Each drift was oriented on a bearing of 299°, to the nearest degree – a direct line on Westminster Bridge. Although the elevation and direction of the guns could not be changed, it would have been possible to alter the range by varying the amount of propellant used in each shot. This would have brought much of London within range.

The railway tunnel ran in a straight line for a distance of about 630 metres. Along its west side was an unloading platform which gave access to a number of cross galleries (numbered 3–13 by the Germans), driven at right angles to the main tunnel at intervals of 24 metres. Each gallery was fitted with a 600 mm gauge railway track. On the east side of the tunnel were chambers intended to be used as store rooms, offices and quarters for the garrison. Trains would have entered the facility and unloaded shells and propellant for the guns.

Galleries 6–10, the central group, gave access to the guns, while galleries 3–5 and 11–13 were intended for use as access tunnels and perhaps also storage areas. They were all connected by Gallery 2, which ran parallel to the main railway tunnel at a distance of 100 metres. Galleries 6–10 were additionally connected by a second passageway, designated Gallery 1, running parallel to the main tunnel at a distance of 24.5 metres. Further workings existed at depths of 62 metres , 47 metres and 30 metres, each serving different purposes associated with the drifts and the guns. The 62 metre workings were constructed to facilitate the removal of spoil from the drifts, while those at 47 metres were connected with the handling of exhaust gases from the guns and those at 30 metres gave access to the breeches of the guns. The lower levels of the workings were accessed via lift shafts, and mining cages were used during construction.

The construction work was carried out by over 5,000 workers, mostly German engineers drafted in from several companies including Mannesmann, Gute Hoffnungshütte, Krupp and the Vereinigte Stahlwerke, supplemented by 430 miners recruited from the Ruhr and Soviet prisoners of war who were used as slave labourers. The intensive Allied bombing campaign caused delays, but construction work continued at a high pace underground. The original plans had envisaged having the first battery of five guns ready by March 1944 and the full complement of 25 guns by 1 October 1944, but these target dates were not met.

In 1943 French agents reported that the Germans were planning to mount an offensive against the United Kingdom that would involve the use of secret weapons resembling giant mortars sunk in the ground and served by rail links. The first signs of abnormal activity at Mimoyecques were spotted by analysts at the Allied Central Interpretation Unit in September 1943, when aerial reconnaissance revealed that the Germans were building railway loops leading to the tunnels into the eastern and western sites. Further reconnaissance flights in October 1943 photographed large-scale activity around the tunnels. An analyst named André Kenny discovered a series of shafts when he saw from a reconnaissance photograph that a haystack concealing one of them had disintegrated, perhaps through the effects of a gale, revealing the entrance, a windlass and pulley. The purpose of the site was unclear, but it was thought to be some kind of shelter for launching rockets or flying bombs. An MI6 agent reported that *'...a concrete chamber was to be built near one of the tunnels for the installation of a tube, 40 to 50 metres long, which he referred to as a 'rocket launching cannon''*. The shafts were interpreted as *'...air holes to allow for the expansion of the gases released by the explosion of the launching charge."* The Allies were unaware of the HDP gun and therefore of the Mimoyecques site's true purpose. Allied intelligence believed at the time that the V-2 rocket had to be launched from tubes or "projectors", so it was assumed that the inclined shafts at Mimoyecques were intended to house such devices.

The lack of intelligence on Mimoyecques was frustrating for those involved in Operation Crossbow, the Allied effort to counter the V-weapons. On 21 March

An artist's impression of the eastern site at Mimoyecques.

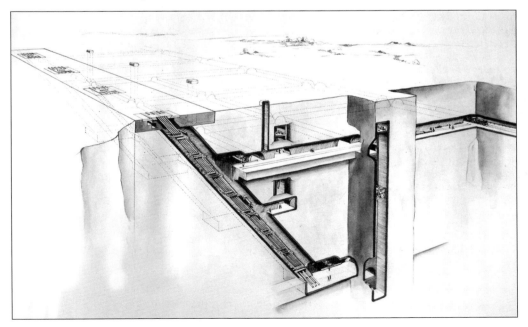

1944 the British Chiefs of Staff discussed the shortage of intelligence but were told by Professor Jones, that little information was leaking out because the workforce was predominantly German. The Committee's head, Duncan Sandys, pressed for greater efforts and proposed that the Special Operations Executive be tasked to kidnap a German technician who could be interrogated for information. The suggestion was approved, but was never put into effect. In the end the Chiefs of Staff instructed General Eisenhower to begin intensive attacks on the so-called 'Heavy Crossbow' sites, including Mimoyecques, which was still believed to be intended for use as a rocket-launching site.

The Allied air forces carried out several bombing raids on Mimoyecques between November 1943 and June 1944 but caused little damage. The bombing disrupted the construction project and the initial raids of 5 and 8 November 1943 caused work to be delayed for about a month. The Germans subsequently decided to abandon the western site, where work had not progressed very far, and concentrated on the eastern site. On 6 July 1944 the Royal Air Force began bombing the site with ground-penetrating Tallboy bombs. One Tallboy hit the concrete slab on top of Drift IV, collapsing the drift. Three others penetrated the tunnels below and substantially damaged the facility, causing several of the galleries to collapse in places. Around 300 Germans and forced labourers were buried alive by the collapses. Adding to the Germans' difficulties, major technical problems were discovered with the HDP gun projectiles. They had been designed to exit the barrels at a speed of about 1,500 metres per second, but the Germans found that a design fault caused the projectiles to begin 'tumbling' in flight at speeds above 1,000 metres per second, causing them to fall well short of the target. This was not discovered until over 20,000 projectiles had already been manufactured.

After the devastating raid of 6 July, the Germans held a high-level meeting on the site's future at which Hitler ordered major changes to the site's development. On 12 July 1944 he signed an order instructing that only five HDP guns were to be installed in a single drift. The two others were to be reused to house a pair of Krupp K5 artillery pieces, reamed out to a smooth bore with a diameter of 310 millimetres, which were to use a new type of long-range rocket-propelled shell. A pair of Rheinbote missile launchers was to be installed at the tunnel entrances. These plans were soon abandoned as Allied ground forces advanced towards Mimoyecques, and on 30 July the Organisation Todt engineers were ordered to end construction work.

The Allies were unaware of this and mounted further attacks on the site as part of the United States Army Air Forces experimental Operation Aphrodite, involving radio-controlled B-24 Liberators packed with explosives. Two such attacks were mounted but failed; in the second such attack, on 12 August, Lt Joseph P. Kennedy, Jr. – the elder brother of future US President John F. Kennedy – was killed when the drone aircraft exploded prematurely. By the end of the bombing campaign, over 4,100 tons of bombs had been dropped on Mimoyecques, more than on any other V-weapons site.

The Mimoyecques site was never formally abandoned, but German forces left it at the start of September 1944 as the Allies advanced north-east from Normandy towards the Pas-de-Calais. It was captured on 5 September by the Canadian 3rd Infantry Division.

Of course, the whole supergun concept surfaced again in 1990 when Canadian engineer Gerald Vincent Bull (*b.* 9 March 1928 – *d.* 22 March 1990) developed a huge artillery piece, known as the Project Babylon 'supergun' for Saddam Hussein and the Iraqi government. Bull was assassinated outside his apartment in Brussels, Belgium in March 1990.

Chapter Five
The RAE Post-War

Before the end of the war it had been realised by all the warring factions that the piston engine had almost reached the zenith of its operational development, and that the turbojet, while still in its operational infancy, would, without doubt become the power plant of the future. By May 1945 both the Allies and Axis powers had jet aircraft in operational service for nearly a year, the Germans with the Arado Ar 234 and Messerschmitt Me.262 and Britain with the Meteor.

In terms of aerodynamic development, the Germans were most greatly advanced for whereas the Meteor and Vampire, and in the United States the P-80 Shooting Star and P-59 Airacomet were of conventional airframe design with so-called 'straight' wings, the Germans had understood the advantages of swept back wings, with a resultantly higher Mach number, and had incorporated these on the Me.262, Me.163, Me.263 and Gotha.229.

When Germany capitulated, great efforts were made by all of the Allies to obtain as much information of these aerodynamic and engine developments as possible, and to this purpose the Allies lost no time in obtaining examples of the latest German jet aircraft and also interviewing the pilots who flew them and the engineers that designed them.

In the late Spring and early Summer of 1945, examples of the latest jet and rocket fighters were ferried to England, the United States - and no doubt Russia - the most common amongst them being the Ar.234, the Me.262, the Me 163 and Heinkel's 90-day wonder the He.162 Volksjager.

According to surviving files, the Me.262 and the He.162 seemed to be of particular interest, mainly because the Allies wanted to learn all they could about both designs, for they feared that information on both aircraft had been passed to the Japanese and they could shortly come up against Japanese versions in the Pacific.

The origins of the He.162 lay in an RLM specification of 8 September 1944 calling for an aircraft with a top speed of 750 kph, an endurance of not less than 20 minutes at sea level, powered by a single BMW003 turbojet, making use of non-strategic materials, capable of being built by semi-skilled labour and to be ready for mass production by 1 January 1945.

In essence what the RLM were asking for was a cheaper complimentary aircraft to the Me.262, which if produced in sufficient numbers could help to turn the tide in the, by then, desperate air war situation over Germany. On 30 September 1944 Heinkel's 162 was adjudged the winning submission, after a close run contest with the Blohm and Voss P211, and Heinkel was given the deadline of 30 October for all development work to be completed, and also a contract for the production of 1,000 aircraft per month.

After a monumental effort by Heinkel the first He.162 *Spatz* (Sparrow) flew on 6 December 1944 from Schweschat airfield. Although the first prototype crashed on 10 December 1944, killing the pilot, the momentum of production was such that throughout December, January and February some 30 prototype and pre-production aircraft were engaged in flight trials. These were mainly aimed at curing the aircraft's instability problems, solved by the fitting of downward turning wing tips and enlarged horizontal tail surfaces.

Plans for putting the He.162 into operational service were commenced on 6 February 1945 when 1/JG1, operating Fw.190s on the Eastern Front, was ordered to move to Parchim and re-equip with the He.162. By the end of April both 2/JG1 and 3/JG1 were also equipped with the aircraft and on 14 April 1945 1/JG1 had been declared operational, although a shortage of fuel restricted operations to a minimum. The 3 May saw all He.162 units based at Leek/Holstein on the Danish border, and on 8 May, by which time the type had achieved but two aerial victories, fifty aircraft were lined up intact ready to surrender to the British forces.

Those German aircraft that were ferried to the UK came under the aegis of the RAE, who took a leading part in the post-war trials of German aircraft in the United Kingdom. This was carried out in three phases:

Phase One - the collection of aircraft from continental Europe, from May 1945 onwards, culminating in the 'German Aircraft Exhibition' at Farnborough in October/November 1945.

Phase Two - an initial evaluation period, commencing with the arrival of the first Luftwaffe aircraft at Farnborough, during which brief handling trials were conducted by Farnborough test pilots on a number of types, together with more extensive engineering assessments and performance measurements of technically innovative types (mainly the jets); during this phase also some were used for routine communications and transport duties.

Phase Three - from 1946 to 1948, the extended aerodynamic assessment of a number of types for reasons of empirical research such as flight trials of the Messerschmitt Me.163s, Horten Ho.IV, etc.

The Ju.287 was intended to provide the Luftwaffe with a bomber that could avoid interception by outrunning enemy fighters. The swept-forward wing was suggested by the project's head designer, Dr. Hans Wocke, as a way of providing extra lift at low airspeeds - necessary due to the poor responsiveness of early turbojets at the vulnerable times of take-off and landing. A further structural advantage of the forward-swept wing was that it would allow for a single massive weapons bay forward of the main wing-spar. [both Hugh Jampton Collection]

The Junkers factory building the V2 and V3 was overrun by the Red Army in late April 1945; at that time, the V2 was 80% complete, and construction of the V3 had just begun. Dr. Hans Wocke and his staff, along with the two incomplete prototypes, were taken to the Soviet Union, where it was discovered to have been built largely from scavenged aircraft parts, including B-24 Liberator nosewheels! Once in Russia, the third prototype (returned to its original Junkers in-house designation, EF 131) was eventually finished and flown on 23 May 1947, but by that time, jet development had already overtaken the Ju.287.

Enemy equipment was abandoned everywhere, some destroyed, such as these Ju.87 Stuka dive bombers seen in the background and some, like this Messerschmitt Me.262, camouflaged under branches. These were discovered near Innsbruck in Austria in May 1945.*[Hugh Jampton Collection]*

There were also plans for a fourth phase - that of either completing, or building from scratch, a number of types of German aircraft for which parts or design details had been discovered by the Allies. The Ministry of Aircraft Production (MAP) listed eighteen types of aircraft which might be assembled in Germany under British supervision. These types were listed in order of priority (with annotations) as follows:

DFS.346	(in US hands, to go to USSR) This was an advanced swept-wing successor to the DFS 228; the '346 was constructed in the USSR after the war.
Bachem Natter	(one complete at Wolf Hirth works, Nabern. Two nearly complete at DFS St Leonards)
Horten Ho.229	(one removed from Gotha works)
DFS.228	(one at Ainring, to go to USSR)
Junkers Ju.287	(in US hands, to go to USSR). This was notable as the first jet aircraft with swept-forward wings. Two Ju.287s were flown by the Soviets after the war.
Messerschmitt Bf.109S	(Boundary-layer control research aircraft) This was an experimental prototype.
Bv.155	(at Finkenwerder, promised to the USA)
Fieseler Fi.103	(turbo-jet version)
Junkers Ju.248	(in US hands, to go to USSR) This was previously designated the Me 163D and was a development of the Me 163 with a retractable tricycle undercarriage.
Tank Ta.183	This was an advanced jet fighter design, with a T-tail and 40-degree wing sweepback.
Messerschmitt P.1101 / Henschell Hs.132	A jet powered dive-bomber of similar layout to the Heinkel He.162 jet fighter.
Junkers EF.126 -	A project, of similar layout to the He.162, for a pulse-jet powered fighter.
Dornier Do.335	(version with jet rear engine). This was the proposed Dornier/Heinkel 535.
Messerschmitt Me 264.	This was a conventional four-engined bomber prototype; the proposed example to have been constructed under British supervision was

	probably a version to have been powered by an experimental aircraft steam turbine. No details appear to have survived of these designs making progress under British auspices.
Dornier Do.635 -	A proposal comprising two Do.335s joined by a common wing centre section.
Hutter Hü.211	The Hutter Hü.211 was a proposed high-altitude reconnaissance aircraft contrived by adding a new wing to a Heinkel He 219 fuselage; the partly complete hydraulic press tool to make the sparless, laminated plywood wing for the prototype was found at Lindengarten, near the Schempp-Hirth works at Kircheim.

The proposal to complete and fly these aircraft was abandoned due to lack of funds and personnel, but it is interesting to speculate on what might have been!

Finally, the BV-238 was a very large flying boat powered by six piston engines, for which an incomplete fuselage and other parts were found at Hamburg. Although they were not included in the original list, significant effort was addressed to completing various Horten tailless designs,

Large, heavily protected underground factories were constructed to take up production of the Me.262, safe from bomb attacks, but the war ended before they could be completed. At B8 Bergkristall-Esche II at St. Georgen/Gusen, Austria, forced laborers of Concentration Camp Gusen II produced fully equipped fuselages for the Me.262 at a monthly rate of 450 units on large assembly lines from early 1945.
[both Hugh Jampton Collection]

The Bachem Ba.349 Natter, or grass-snake, was a World War II German point-defence rocket-powered interceptor, which was to be used in a very similar way to a manned surface-to-air missile. After a vertical take-off, which eliminated the need for airfields, most of the flight to the Allied bombers was to be controlled by an autopilot. The primary role of the relatively untrained pilot was to aim the aircraft at its target bomber and fire its armament of rockets. The pilot and the fuselage containing the rocket-motor would then land using separate parachutes, while the nose section was disposable. The only manned vertical take-off flight (left) on 1 March 1945 ended in the death of the test pilot, Lothar Sieber.

French forces captured Waldsee by 25 April 1945 and presumably took control of the Bachem-Werk. Just before the French troops arrived, a group of Bachem-Werk personnel set out for Austria with five A1 Natters on trailers. They were captured by US troops at the junction of the river Inn and one of its tributaries, the Ötztaler Ache, at Camp Schlatt around 4 May. *[all Hugh Jampton Collection]*

including Horten Ho VII training aircraft found in the Peschke works in Minden and the larger Horten Ho VIII transport prototype, WNr4l, which was found three-quarters complete at Minden and was moved to Gottingen to be completed in the original Horten factory. In the end these efforts came to nothing.

The main thrust for the proposal to complete these aircraft came from the Air Ministry. It was proposed that the work would be centred on the experimental establishment at Volkenrode, the former Luftforschungsanstalt (LFA) Hermann Goering, which had been taken over as RAF Volkenrode, under the command of Wing Commander Geoffrey Mungo Buxton. The proposal to complete the aircraft was abandoned due to lack of funds and resources. Instead effort was spent in studying the material found at the Establishment and in dismantling major items of equipment, such as wind tunnels, for

transport to the UK to be used at British research establishments. The main product in terms of hardware was a quantity of Walter 109-509 rocket engines manufactured after the end of the War by the Walter factory, to be tested at the RAE Farnborough.

Phase One
The collection of ex-Luftwaffe aircraft for evaluation had been started by the British Air Ministry's Branch AI-2(g). This organisation had been established in 1942 as a research unit to undertake research into new enemy aircraft to anticipate the introduction of new equipment. AI-2, with assistance from the British Wartime Ministry of Aircraft Production, drew up a 'Requirements List' of items needed for evaluation in the UK after the war. AI-2(g) deployed inspection teams in all theatres of war to examine downed aircraft and established an operations centre within the Air Ministry to coordinate new information for all sources, including interviewing prisoners of war.

This 'wants list' had been initiated during 1944 and was distributed to the Air Technical Intelligence teams in Europe prior to the German collapse in May 1945. The list was amended as new requirements were identified; including the addition of previously unknown aircraft or items of equipment found on the ground by the intelligence teams. With the end of hostilities the Air Technical Intelligence teams were reinforced by experienced pilots and engineers, many of them from the RAF Central Fighter Establishment at Tangmere. The personnel included members of the former 1426 Flight which had been incorporated into the CFE shortly before the end of the war.

It was decided that the ferrying of unfamiliar aircraft types was best carried out by trained test pilots and so the whole process of selecting and ferrying German aircraft to the UK was handed over to the RAE. Lt Cdr Eric

Dornier D.217M 'AM 107' was captured at Beldringe in Denmark. and arrived at Farnborough on 13 October 1945. *[Hugh Jampton Collection]*

Heinkel He.162A-2 'AM 62' was displayed in Hyde Park, London during Battle of Britain Week in September 1945. After a period of storage at Brize Norton it was shipped to Canada. *[Hugh Jampton Collection]*

Melrose 'Winkle' Brown was placed in charge of the reception of German aircraft at Farnborough, and his superior, Group Captain Alan F. Hards, the Commanding Officer Experimental Flying at RAE, took over responsibility for the selection of suitable aircraft. The servicing of aircraft prior to their delivery to Britain remained the responsibility of the RAF and this task was carried out by 409 Repair and Salvage Unit, based at Schleswig in northern Germany. The RAE set up an outpost at Schleswig, commanded by Squadron Leader Joe McCarthy, to co-ordinate the delivery of selected aircraft to Schleswig for overhaul, and to control the acceptance test flights of individual aircraft at the completion of their servicing routine. The RAE then took over the delivery of the aircraft to England via one or more staging posts in Holland or Belgium which were provided with jet fuel and other support facilities.

The collection phase started on 18 May 1945 when Wing Commander Roland John 'Roly' Falk flew the Messerschmitt Me.262B night fighter 'Air Min 50' from Schleswig to Gilze-Rijen, continuing onwards to Farnborough the next day. The exercise was completed on 18 January 1946 when Focke

Above: taken on 2 October 1945 in front of 'A' Shed at the RAE Farnborough, this picture shows the *Hirschgewihe* aerials and the underwing fuel tanks of Bf.110G-4/R3, Air Ministry 30.

Right: A publicity shot taken at Grove, showing the 'Air Ministry' number being painted on Bf.110G AM 30.

Junkers Ju.88AA 'AM 108' seen at RAE Farnborough, with the later Ju.88M seen behind it.

Wulf Fw.58 insecticide-sprayer 'Air Min 117' arrived at Farnborough from Schleswig, via Gütersloh and Melsbroek. In the meantime approximately seventy-five ex-Luftwaffe aircraft had made the journey from Schleswig to Farnborough. Nearly fifty others had arrived there by surface transport and a few more had flown in from other places. The whole operation was safely conducted with only two airframe losses - one Ar 234B was damaged beyond repair on landing at Farnborough and one Siebel Fh 104 had force-landed in the sea en route, but there were no crew injuries.

To give an idea as to the scale of the operation, those aircraft arriving by surface transport included twenty-three Me 163s, eleven Heinkel He.162s and the Messerschmitt Me.262C prototype. Others involved in the move by surface means included a prototype Blohm und Voss Bv.155B high-altitude fighter and the first prototype DFS.228 high-altitude reconnaissance aircraft. These last two aircraft were loaned to the British by the US authorities for investigation at Farnborough before their intended shipment to the USA.

Following their arrival at Farnborough, very many of the aircraft were flown or taken by road to the RAF's 6 Maintenance Unit at Brize Norton in Oxfordshire to be stored, pending allocation to a series of tests.

The DFS.346 - the first item on the MAP list - was a rocket-powered swept-wing vehicle subsequently completed and flown (with indifferent success) in the Soviet Union after the war. It was designed by Felix Kracht at the *Deutsche Forschungsanstalt für Segelflug* (DFS), the 'German Institute for Sailplane Flight'. The prototype was still unfinished by the end of the war and was taken to the Soviet Union where it was rebuilt, tested and flown.

The DFS.346 was a parallel project to the DFS.228 high-altitude reconnaissance aircraft, also designed under the direction of Felix Kracht and his team at DFS. While the DFS.228 was essentially of conventional sailplane design, the DFS.346 had highly-swept wings and a highly streamlined fuselage that its designers hoped would enable it to break the sound barrier.

Heinkel He.162A 'AM 61' seen at Farnborough.

Like its stablemate, it also featured a self-contained escape module for the pilot, a feature originally designed for the DFS 54 prior to the war. The pilot was to fly the machine from a prone position, a feature decided from experience with the first DFS 228 prototype. This was mainly because of the smaller cross-sectional area and easier sealing of the pressurised cabin, but it was also known to help with g-force handling.

The 346 was intended to be air-launched from the back of a large aircraft, the baseline machine being the Dornier Do.217. After launch, the pilot would fire the 346's Walter 509B/C rocket engine to accelerate to a proposed speed of Mach 2.6 and altitude of 100,000 feet. This engine had two chambers, high and low thrust, and after reaching altitude the speed could be maintained by short bursts of the smaller chamber.

In operational use the aircraft would then glide over England for a photo-reconnaissance run, descending as it flew but still at a high speed. After the run was complete the engine would be briefly turned on again, to raise the altitude for a long low-speed glide back to a base in Germany or northern France. There is strong evidence that the work done by DFS on these high altitude powered sailplanes formed the basis of the development of the Lockheed U-2 spy-plane of the 1950s in the USA.

In September, a number of German aircraft had been displayed in Hyde Park in London, during what was termed a 'Thanksgiving Week'. These aircraft included the Fieseler Fi.156 Air Min 100, Messerschmitt Bf.108 Air Min 84, Messerschmitt Me.163B WNr 191454, Messerschmitt Bf.110G AM 85, Junkers Ju.88G-6 Air Min 3, and Heinkel He.162 Air Min 62 / WNr

He.162 fuselages being constructed on their jigs within Heinkel's southern underground facility at Hinterbruehl near Moedling, in the spring of 1945. (USAAF)

This unidentifiable Me.262A was photographed at Twente, in the Netherlands prior to reportedly being delivered to Farnborough.

120086 and a Grunau Baby IIB glider.

The collection of the many and varied types enabled the RAE to mount a 'German Aircraft Exhibition', held at Farnborough from 29 October to 9 November. The Exhibition included displays of aircraft and associated displays of equipment in various hangars and workshops, a static display of some of the larger aircraft on the airfield, and, on some days, a flying display of a few of the more interesting types, including the jets. For this Exhibition, several aircraft were recalled from storage at Brize Norton to be shown to the public.

The external static display included the following:
Arado Ar.232B Air Min 17
Junkers Ju.290A Air Min 57
Junkers Ju.52/3m Air Min 104
Junkers Ju.352 Air Min 109
Focke Wulf Fw.200C Air Min 94
'Mistel' composite Air Min 77 (Ju.88A WNr 2492 A + Fw.190A W Nr 733759)
Dornier Do.217M Air Min 106
Siebel Si.204D Air Min 4
Junkers Ju.88G TP190/Air Min 231
Reichenburg IV (piloted Fieseler Fi.103/ V1)
Junkers Ju.188A Air Min 108 / W Nr 230776

Me.163B W.Nr 191454 Air Min 204, seen in London's Hyde Park during September 1945.

Junkers Ju.3881. Air Min 83
Heinkel He.219A Air Min 22 / W Nr310189
Focke Wulf Fw.189A Air Min 27 / W Nr 0173
Messerschrnitt Me.410B Air Min V-3 (Air Min 74)
Messerschmitt Bf.110G Air Min 30 / WNr 730037
Dornier Do.335A Air Min 223
Tank Ta.152H Air Min 11 / WNr 150168
Focke Wulf Fw.19013 Air Min 111
Messerschmitt Bf.109G VD358
Messerschmitt Bf.108B Air Min 84
Fieseler Fi.156C Air Min 100
Arado Ar.234B VK877 (Air Min 26) / WNr 140476
Messerschmitt Me.262A Air Min 80 / WNr 1 1 1690
Heinkel He.111H WNr 701152 (arrived on 3rd November)
Focke Wulf Fw.190D WNr 210079

The Reichenburg was almost certainly captured at the Dannenburg V1 factory in the US Zone. The fate of this aircraft is unknown; one of the surviving V-1s in England is a standard pilotless aircraft modified to appear like a piloted version.

The Fw.190D had been shot down in Belgium during 'Operation Bodenplatte' on New Year's Day 1945 and its remains had been taken to Farnborough for examination. It was in a derelict condition and was available for children to climb on.

The main flying display took place on 4 November and comprised of short flypasts by Dornier Do.217 Air Min 107, Focke Wulf Fw.190A PN999, Bücker Bü.181 Air Min 122, Messerschmitt Bf.108 Air Min 89, Junkers Ju.52/3m Air Min 103, Fieseler Fi.156C Air Min 99, and Junkers Ju.88 Air Min 42/WNr. 622461. The Do.217 and Fw.190 made two separate flights and other aircraft which took off during the day included Heinkel He.162A Air Min 61 and Siebel Si.204 Air Min 28. On the opening day, flypasts had been made by the Heinkel He.162s VH513 and Air Min 61 and by the Messerschmitt Me.262A Air Min 81.

The only unfortunate event during the Exhibition was the crash of Heinkel He.162 Air Min 61 on the closing day. Flt Lt R. A. Marks, the pilot, was killed in the accident together with a soldier in the barracks adjoining the airfield, on which the aircraft crashed.

The DFS.228 (Deutschesforschung sanstalt fur Segelflug - which translates as the 'German Institute for Sailplane Flight') was a rocket-powered, high-altitude reconnaissance aircraft, carried aloft by a Do.217, landing back on skids. The fuselage of the prototype DFS.228V1 seen at Farnborough in 1946. The aircraft carries the German civil registration D-IBFQ.

A variant of the DFS.228 was the DFS.346 (left) which was completed and test flown in the Soviet Union after capture by the Russians.

In 1947, an entirely new DFS.346 prototype was constructed, incorporating refinements suggested by the tests. This was designated DFS.346-P. No provision was made for a powerplant, but ballast was added to simulate the weight of an engine and fuel. Centre: Test pilot Rolf Mödel tries out the prone position.

The DFS.246-P was carried to altitude by a B-29 Superfortress captured in Vladivostok and successfully flown by Wolfgang Ziese in a series of tests.

Phase Two

This involved engineering assessments, followed by limited flight tests. Such work was concentrated mostly on jet types. Typical of the work was that carried out on the Messerschmitt Me.262A WNr.112372 (Air Min 51). This aircraft was flown from Lubeck to Copenhagen/Kastrup, and then on to Farnborough, on 23 June 1945. It was taken over by the RAE Structures and Mechanical Engineering Flight, which carried out a detailed engineering assessment of the aircraft, that later resulted in the publication of Report No. EA 235/3 describing the Me.262's electrical system. The machine was then turned over to the Aerodynamics Flight for flight tests, which were conducted between September and November 1945. The test flights contributed to Report No. EA 235/5 *'Note on performance in flight of German jet propelled aircraft Me.262, He.162, Arado 234'*. This report summarised data obtained during about 30 flying hours on these types.

The Junkers Ju.352s provided facilities not available on any Allied transport aircraft of the period. This was an intended replacement for the well-known Ju.52/3m, featured an hydraulically-operated tail-loading ramp, leading to a cargo hold which was quite spacious by the standards of that era, and had a payload capacity approaching five tons. Five of these aircraft were captured and made many trips between Farnborough and Germany to bring back captured equipment for evaluation or for more long-term use at Farnborough and elsewhere. A few flights were also made by the sole Arado Ar.232B tactical transport to be captured. That design also featured tail-loading directly into a cargo hold. The Siebel Si.204s were busily employed during 1945 on communications duties, often carrying ferry pilots to collect other German aircraft or providing navigational guidance to single-seat types on long ferry flights. The last recorded flight of a Siebel at Farnborough was on 2 January 1946, when 'Air Min 28' departed for storage at RAF Brize Norton.

Shortly after this, on 18 January 1946, the Dornier Do.335A tandem-engined

The Mistel S3A combination - a Ju.88 with Fw.190 on top is seen at Farnborough. The combination had been surrended at Tirstrup in Denmark.

fighter 'Air Min 223' caught fire during take-off and crashed near the airfield, killing Group Captain A. F. Hards, Commanding Officer Experimental Flying. After this tragic accident, very severe restrictions were placed on flying by ex-German aircraft, effectively bringing to an end 'Phase Two'.

Following this accident the only use of ex-Luftwaffe aircraft was as transports, or, as in the case of the use of the Fieseler Storch, various basic aerodynamic investigations. Several examples of the Fieseler Storch were taken to Farnborough and one of these was flown there for ten years after the war. The Storch was used for many different duties, but during 'Phase Two' these included type handling trials and performance assessment.

The longest serving of the transport aircraft was the Junkers Ju.352A 'Air Min 8', which arrived at Farnborough on 23 June 1945 and remained in service until 28 June 1946. Its retirement seems to have been due to serviceability problems. On 27 June 1946, three Junkers Ju.52/3m were delivered to Farnborough after overhaul at Hamburg-Fühlsbuttel. It was intended that these should serve with the RAE General Duties Flight in the transport role. Although these aircraft made a number of flights to Germany, mainly to the research institute at Volkenrode to collect equipment to be used at Farnborough, they did not survive for long because of serviceability problems, the last being retired in October 1946.

Following the cessation of hostilities British airlines - in particular British European Airways and Railway Air Services - were faced with a lack of suitable aircraft and the surplus of surrendered Ju.52/3ms seemed to be the answer. The Ministry of Civil Aviation made arrangements with the RAF for 40 aircraft to be flown to the UK (including ten to be cannibalised for spares) with additional aircraft reserved for possible use by BOAC. 6 MU was given the task of storage, pending conversion or breaking for spares and Field Consolidated Aircraft Services Ltd was given the contract to recover spares

Part of the aircraft exhibition at Farnborough. Aircraft on display include a Bf.109G, Fw.190A, Focke Achgelis Fa.330 and a pair of He.162s. Suspended in the background was the DFS108-80 sailplane. A number of these aircraft had their panels open to show internal structure or had been partially sectioned.

from cannibalised aircraft. In the event more than 50 Ju 52/3ms were delivered to Brize between November 1945 and June 1946. Fifteen were passed on to Short Brothers and Harland at Belfast for civilian conversion but the remainder lingered on for a couple of years until being struck of charge on 11 March 1948 and sold to J Dale & Co Ltd for scrap. One Ju.52/3m did not make it to Brize, having been damaged en-route at Manston in June 1946 and scrapped. Another, after leaving for Belfast on 21 March 1946, force landed at Little Rissington, requiring a working party to carry out repairs allowing it to continue on its way over a month later. VM980 which was flown to Brize Norton in February 1946, was formerly SX-ACF with Hellenic Airways and had been taken over by the Luftwaffe during the invasion of Greece in April 1941. Other aircraft were flown direct to Belfast after this time, bypassing 6 MU.

Phase Three

After the accident to the Dornier Do.335, an overall reassessment of the use of the ex-Axis aircraft was made, resulting in a period when only the Fieseler Fi.156 and one of the Junkers Ju.352A transports were flown. Then, in March 1946, flight trials of the Messerschmitt Me.163B VF241 restarted. These trials were made mostly at the nearby airfield of Wisley, to avoid the busy circuit traffic at Farnborough, since the Me.163B was towed off as a glider by a Spitfire and released at altitude to make its own way back to earth. These trials were primarily to explore the handling characteristics of the Me 163B's tailless configuration, to provide information for other tailless designs on the drawing boards of British manufacturers in the post-war period. The 'tailless' design concept had been attracting aircraft designers for years, as evidenced by the experimental designs of Northrop in the USA, Lippisch and the Horten brothers in Germany, and innovative designs by Handley-Page and Miles Aircraft in the United Kingdom. During the war years, the British Ministry of Aircraft Production had placed several contracts for experimental tailless aircraft with the aircraft industry. The tests of the Me.163B, and of the Horten Ho.IV sailplane, were conducted as part of this overall programme. In April 1946 decisions were taken about the long-term

The 'Internal Aircraft Exibition' at Farnborough was not so much that the aircraft were inside, but related more to the aircraft's internal structure! Some fascinating aircraft were on display, including the Horten Ho.IV flying wing glider.

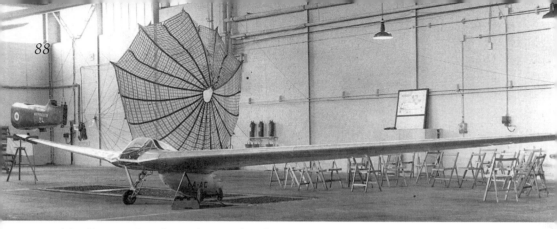

The Horten Ho.IV 'LA-AC' seen in 'A' Shed at Farnborough. Above and behind it is an example of the German 'ribbon' parachute.

use of the German aircraft, resulting in the allocation of RAF serial numbers to those which it was intended to maintain in flying condition. These aircraft were: VP543 Horten Ho.IV sailplane VP546, Fieseler Fi.156 Storch VP550, Junkers Ju.352A VP554, Messerschmitt Me.262A VP559, DFS.108-14 Schulgleiter SG.38 VP582, DFS.108-14 Schulgleiter SG.38 VP587, DFS.108-49 Grunau Baby VP591 and DFS.108-30 Kranich II sailplane.

In addition to these, the Messerschmitt Me.163B had previously been given an RAF serial number, VF241. It was flown initially to provide general experience in the handling of tailless aircraft and later to accumulate experience in the skid landing technique. The Me 163 was launched on a wheeled trolley which was jettisoned after take off. Landings were made on a hydraulically-sprung skid which was lowered prior to landing. Interest in this technique arose because two German aircraft designers, Dr Martin Winter and Dr Hans Multhopp, were employed at the RAE for four years post war, designing a number of high-speed experimental jet aircraft including at least one which was to land on skids. Also at this time a naval jet fighter programme was under development based on the presumption that, to save weight, the aircraft would not have a conventional undercarriage, but would land on a rubberised 'flexible deck' on the aircraft carrier. After the trials of the Me 163B, which ended when the aircraft was damaged beyond repair following a high-speed skid landing at Wittering, interest in the skid-landing concept waned.

Born in 1913, Multhopp studied at the Technische Hochschule in Hannover, before transferring to the University of Göttingen in 1934. Studying there under the guidance of the famed aerodynamicist Ludwig Prandtl, Multhopp assisted in the design and construction of several gliders while working on a thesis on the subject of wing aerodynamics. During his time at the university, he also participated in experiments under the aegis of the *Aerodynamische Versuchsanstalt* (AVA), the German counterpart to the

The Blöhm und Voss Bv.222V-2 seen at Trondheim in 1945 wearing RAF markings. This aircraft was later painted with large 'stars and stripes' insignia and a US Navy crew flew to pick it up to ferry it to the USA for intelligence gathering, but the ferry flight never happened and the six-engined flying boat was scuttled at Trondheim. *(USAAF)*

This Dornier Do.335 wearing US markings sits next to the burnt out wreckage of at least one Me.262 and one He.162. *(USAAF)*

Blöhm und Voss Bv.222C-012 marked as VP501 and in full RAF markings but with two of the starboard engines missing. The flying boat is seen at RAF Calshot during 1946 *(USAAF)*

American National Advisory Committee for Aeronautics. In 1937, Multhopp was placed in charge of one of AVA's wind tunnels. More importantly, he had already published a seminal paper on wing-lift theory and before he had fulfilled his doctoral requirements, his work had attracted the interest of other German aviation concerns. Consequently, Multhopp was approached by Kurt Tank in 1938, who offered him employment at Focke-Wulf.

There Multhopp was promoted to assistant in charge of the aerodynamics department in 1940, then to advanced design bureau chief in 1943. One of his innovative projects during this time was the Multhopp-Klappe, a combined flap and dive brake apparatus, which was installed on the unsuccessful Focke-Wulf Fw 191. During flight testing of the Fw 191, when deployed, the Multhopp-Klappe caused severe flutter.

In 1944, in collaboration with Tank and his design team, Multhopp was assigned to a research project to meet and exceed the specifications of the Reichsluftfahrtministerium (RLM)'s Emergency Fighter Competition for a single-seat jet-powered fighter intended for performance at high-altitude. His design of the diminutive Ta 183 Huckebein distinguished by a 40° swept-wing and T-tail was the winner of the Luftwaffe's 1945 Emergency Fighter Program. However, due to the deteriorating war situation, delays in development meant that only wind tunnel models had been completed by the time of Germany's surrender.

Although some historians claim that the post-war Soviet MiG-15 was

based on the Ta.183, modern experts in Russian and Soviet aviation history reject this claim, although it is acknowledged that some of the captured data from Multhopp's design work was examined by Artem Mikoyan and Mikhail Gurevich in the formative study of contemporary research. The swept-wing data that was amassed at Focke Wulf was, however, utilized by the Saab design office in its preliminary work that led to the Saab J29 fighter. A member of the Saab engineering team had been allowed to review German aeronautical documents stored in Switzerland. These files, captured by the Americans in 1945, clearly indicated delta and swept-wing designs had the effect of '...reducing drag dramatically as the aircraft approached the sound barrier.' Although more sophisticated than the Ta.183, the Saab J.29 had more than a superficial link to the earlier German fighter project.

Emigrating to the United Kingdom after the war, Multhopp assisted in the advancement of British aeronautic science before moving to the United States, where his work for Martin Marietta on lifting bodies provided aerodynamic experience that proved instrumental in the development of the Space Shuttle.

The Multhopp project was never built, due to a government policy decision not to risk pilots' lives in supersonic flight, a decision which later cost Britain its lead in jet fighter development.

The Ho.IV made its first flight on 27 March 1946, and again on 18 April. Thereafter there is no record of it making further flights until a new series of nine flights was made in April, May and June 1947. It seems possible, however, that the official Flight Log does not contain a complete record of the flights made by gliders. Most of the Ho.IV flights are believed to have been made by Robert Kronfeld, who had been instrumental in bringing the sailplane to England in the first place, and after the Ho.IV was withdrawn from use at Farnborough, it was sold to him.

The Fi.156, VP546, was maintained in flying condition at Farnborough until 1955, when it was grounded, due to lack of spare parts. It was used for a large variety of different projects. These included aircraft-carrier deck landings (on HMS *Triumph* in 1946, flown by 'Winkle' Brown), formation flying with helicopters to allow air-to-air photography of rotor blade behaviour, glider-towing, and routine communications flying. In 1948, another Fi.156, which had been used by AV-M H. E. Broadhurst as a personal transport, was added to the RAE fleet (serial number VX154). This was flown for two years on similar duties, until it was grounded to act as a source of spares for the original aircraft.

The Ju.352, VP550, only made two return trips to Germany after April 1946, before it was grounded due to servicing problems, while the proposed trials with the Messerschmitt Me.262 VP554 were cancelled and the aircraft was shipped to Australia.

The remaining aircraft - all of them gliders - were taken on charge for use by the RAE Technical College Glider Flight. Their main duty was to give elementary flight training to the apprentice engineers under training at the Royal Aircraft Establishment's own training establishment. These aircraft were also used to give glider experience to RAE test pilots who might have

The Arado Ar.232B-0 tactical transport seen as 'Air Min 17' at RAE Farnborough. The aircraft had been operated by III/KG 200 as A3+RB and was surrendered at Eggebeck.

Junkers Ju.352 'Air Min 8' seen in Germany on one of its many transport shuttle missions.

to conduct trials with unpowered aircraft. In later years the two-seat Kranich was fitted with blind-flying instruments and two-stage amber screens in one cockpit to conduct a series of simulated blind-flying experiments, in conjunction with the Empire Test Pilot School, which was also based at Farnborough at the time.

The only other significant trials of an ex-German aircraft took place in 1947. At that time, there was some concern at the level of light aircraft accidents, and also pressure from some quarters to see the possibility of pilot training carried out on single-seat aircraft. The authorities were giving consideration to this being allowed, providing that the aircraft concerned was 'unstallable'. A series of trials were therefore carried out at Farnborough, involving the Brunswick Zaunkonig V-2 light aircraft, which had been built at the Brunswick Technical High School in 1944. This design featured a wing with full-span slots and flaps and was intended to be unstallable. The prototype was brought to Farnborough and was first flown there on 18 September 1947. After a long series of flights in 1947 and early 1948, the Zaunkonig was handed over to the Civil Aviation Flying Unit of the (then) Ministry of Civil Aviation on 16 July 1948. This

Two views of a German Junkers Ju.290 marked as '022' on one of its fins, captured by the Americans at Orly Airport, Paris.

The Brunswick Zaunkonig V-2 light aircraft was allocated RAF serial VX190, and later appeared on the British civil aircraft register as G-ALUA.

unit, based at Gatwick Airport near London, carried out further trials. In March 1949 the aircraft was returned to the RAE and then sold in May 1949 to the Ultra Light Aircraft Association (ancestor of the present-day Light Aircraft Association).

Systems of Identification.
British aircraft serials were allotted primarily to those aircraft which were allocated to test and evaluation flying programmes. Serials were issued as and when required, in the normal current sequences, the first being AE479 to a Bf 109E-3, in June 1940, and the final allotment VX190 to the Brunswick Zaunkonig, in about 1946.

In 1945-46, when the large influx of captured enemy material arrived for technical evaluation, the Air Ministry allocated numbers between 1 and 250, with the prefix Air Min, or just AM, to specific airframes.

So many aircraft were received, that space soon became a premium. 6 Maintenance Unit (MU) was selected as the storage unit for German aircraft which had come into RAF hands. These aircraft were stored on behalf of the RAE, who had evaluated many of them, from where they were flown or roaded in. An additional 98 service personnel were allocated to the unit for this commitment. The first aircraft to arrive was a Ju 188A with RAF serial VG916, which flew in on 30 May 1945, having been flown from Trondheim, Norway to RAF Fraserburgh in Scotland on 2 May by a defecting Luftwaffe crew.

The purpose of these 'Air Ministry' numbers was initially to identify aircraft of intelligence interest at their place of surrender in Germany or Denmark, and to clearly segregate such aircraft from the far larger number of aircraft which were to be destroyed as being of no further interest or use. A typical airfield at the time of the surrender in May 1945 held literally hundreds of Luftwaffe aircraft of which perhaps ten were selected as being 'Category

Heinrich Himler was Reichsführer of the Schutzstaffel (SS), a military commander, and a leading member of the Nazi Party. He had his own personal Fw.200 Condor transort. The aircraft was captured at Flensburg where the Regierungsstaffel - the German goverment's VIP Transport Squadron had moved before the German surrender. The Condor is seen here at Farnborough wearing the 'AIR MIN 44' number.

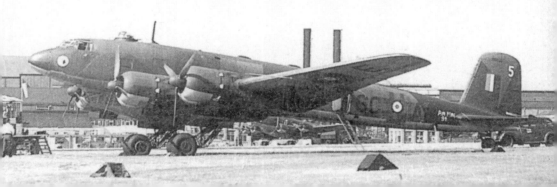

Just because an aircraft was allocated an 'Air Min' number, did not always mean it was visible. This Bf.110G (W Nr 180560) was ferried to the UK from Knokke-le-Zoute and given the number AM 15.

A captured He.162, with another wrecked example alongside.

One' for evaluation in the United Kingdom, and a similar or slightly larger number of communications or trainer types that would be allocated to other categories assigning them to use by the Royal Air Force in Germany or for transfer to Allied governments for use by their own air forces.

One aircraft which did not make it in to Brize Norton was Fw.190S AM37, which crashed at Sonning, north of Reading, killing the pilot, when the prop pitch control failed. It is believed that Me.262A-2A AM 52 was flown by the American Bell company chief test pilot Jack Woolams when it was with the MU. Jet fuel, a comparative rarity at this time, was drained from stored Ar.234Bs for the flight. Woolams also made the 15 min delivery flight of He.162A-2 AM 59 from Farnborough on 2 August 1945. These flights were no doubt connected with Bell's interest in jet aircraft development.

The 'AIR MIN' designation derived from the fact that the original team who travelled around ex-Luftwaffe bases selecting aircraft for evaluation, reported directly to the Air Ministry in London. Its selections were made against a list of requirements which had been compiled by department Air Intelligence 2(g) at the Air Ministry, in consultation with the Ministry of Aircraft Production.

It was not long before both the selection process and the responsibility for ferrying the selected aircraft to the UK was handed over to the RAE, largely because it was decided that it would be safest for RAE test pilots to fly the unfamiliar types of aircraft. Additionally, and for now unknown reasons, the Air Ministry allocated to Me 410 aircraft a series of V numbers, for example, Air Ministry V3 was an Me 410B-6.

As a rule the Air Ministry numbers were crudely applied to the aircraft

in white or black paint on the rear fuselage, aft of the roundel, usually on the port side only, but there were exceptions to this practice. At least one machine, Bf.110G-4/R7, c/n 730037, had the number painted neatly and in full, as AIR MINISTRY 30, for the benefit of official publicity photographs.

Some of the larger aircraft had their identities painted on both sides and, very rarely, under the wings. Most commonly the identity was 'Air Min', although occasionally 'AM', or in full as 'Air Ministry'. The highest known allotment is Air Ministry 231, a Ju.88G-l, but it should be noted that although numbers were allocated to specific aircraft, these numbers were not necessarily applied to the airframe. Just to confuse things even further, an aircraft could be allocated an Air Ministry number but if it was subsequently required for test flying a normal RAF serial allocation was made in addition!

Many aircraft did not receive 'Air Min' numbers at their points of surrender - these included the Messerschmitt Me 163B rocket-powered fighters, which were selected by a separate team of mainly RAE personnel formed solely to investigate that type of aircraft. Other categories which did not at first receive numbers were aircraft selected in the US Zone of Germany, against British requirements identified in the joint US/UK list. The same applied to a number of aircraft surrendered in Norway. The latter area had not been considered of prime intelligence interest and was not visited by the Air Ministry team, but several aircraft were reported by the RAF's 88 Group, which moved to Norway to assist the Norwegians to liberate their country. In fact the intelligence assessment had been almost correct, given that most of the 'Category One' aircraft selected in Norway had flown there from Denmark on the day that German forces in that country surrendered, 5 May 1945.

Aircraft in the three categories arrived at Farnborough and were allocated 'Air Min' numbers in the series 200 onwards. By this time the responsibility for allocating 'Air Min' numbers in the original series had been transferred to the RAE, the selection of suitable aircraft generally being made

Air Ministry 20 - Heinkel He.219A from I/NJG3 and equipped with FuG220 radar was surrendered in Denmark, but did not arrive at Farnborough until 3 August 1945.

Fiesler Fi.156C-7 Storch VT+TD was photographed at Farnborough on 11 November 1945 with the crudely applied 'Air Min 99'.

Heinkel 111H 'FE-1600' seen at Freeman Field in the USA. As with many captured aircraft, mystery surrounds it. The aircraft was flown to San Servio, north of Foggia, Italy by a Hungarian pilot on 14 December 1944. Then, depending on which record is used, the aircraft was either flown to Wright Field USA during January 1945, or shipped there at a later date!

by Group Captain Alan Hards, Commanding Officer Experimental Flying.

Enemy Aircraft numbers were allocated by the Enemy Aircraft Flight, Central Fighter Establishment, Tangmere, early in 1945, when captured enemy aircraft were acquired subsequent to the demise of No. 1426 (EAC) Flight. The letters EA preceded the roundel, which was followed by a number, although very few allocations are known.

Although very little is known about the system, it would appear that the General Duties Flight at the Royal Aircraft Establishment, Farnborough, allocated a series of numbers to specific aircraft operated by them, eg GD1 was a Ju 52/3m.

On 9 June 1945, Group Captain Hards visited Copenhagen/Kastrup and Vaerlose in Denmark and selected three Blohm und Voss Bv 138 flying-boats at Kastrup-See and three Messerschmitt Me 410s at Vaerlose. These aircraft became Air Min 69 to 74 inclusive, although initially they had identities such as 'Air Min S-2' and 'Air Min V-3' painted on them.

The reason for this is not known, but one theory promulgated is that they mark the transition point of allocation by the original 'Air Ministry' team to allocation by the RAE. It is reasonable to think that when he selected the six aircraft on 9 June Group Captain Hards created some identities on the spur of the moment, for he did not know at what point the allocations had been reached in the numerical series. The anomaly could and was corrected prior to the arrival of the Messerschmitts at Farnborough, for although they flew in with numbers such as 'Air Min V-1' painted on them, their Flight Log

Flettner Fl.282 T2-4614 seen after arrival at Pewitt Aircraft of Benedict Airport, Booth Corners, Pennsylvania. It was canibalised for spares to make T2-4613 airworthy. (USAAF)

entries show both identities, as does the inventory of German aircraft received. The 'S' is thought to indicate Kastrup-'See', while the 'V' indicated 'Vaerlose'. At least two aircraft from the Kastrup land airfield had the letter 'K' added to their previously allocated 'Air Min' numbers - AM K41 and K42 – and it is possible that other aircraft at Kastrup also had this, but none came to the UK for this to be noted by observers. It may be relevant that the three aerodromes (Kastrup, Kastrup-See and Vaerlose) were run by the same squadron of the RAF disarmament organisation, 5 Squadron, 8403 Air Disarmament Wing.

By early July 1945 the 'Air Min' numbers had passed the 100 mark, after which allocations tailed off in quantity, the final allocations in Germany being made in September 1945. Those made at Farnborough, from AM 200 onwards, were all made prior to the German Aircraft Exhibition in October 1945.

The USAAF intelligence teams operating in Europe had their own list of requirements, but the British and American lists were soon amalgamated into one. US requirements selected in British controlled areas were initially numbered in a parallel series of identities starting at 'USA 1'.

Nevertheless, after receipt at Wright Field, the US-identities incorporating the letters 'EB-' (Evaluation Branch), 'FE-' (Foreign Equipment) or 'T2-' (for the T-2 Office of Air Force Intelligence) were painted on the aircraft. The identity numbers were unique to the aircraft and were rarely duplicated, although the prefixes EB-, FE- and T2- were valid on the same airframe at different times, as the intelligence organisation underwent changes. It was usual for the prefix to be changed if the aircraft remained on charge after one of the organisational changes -thus several 'EB-' aircraft changed to 'FE-' and later 'FE-' aircraft became 'T2-'. In fact the vast majority of 'FE-' numbered aircraft had their prefix overpainted as 'T2-' which survived into the post-Second World War period, being used again for North Korean aircraft captured during the Korean War (such as the Yak-9P T2-3002).

Demonstrating the point that just as captured aircraft in RAF services did not always carry their 'Air Min' numbers, this Dornier Do.335 was on the strength of the Tactical Test Division at BAS Patuxent River. The '102' on the fin relates to its Werk Nummer, 240102.

Chapter Six
Exploiting the Enemy's Developments

Just as the Royal Aircraft Establishment at Farnborough and other airfields and organisations tested captured enemy aircraft in the UK, the same happened in the USA.

Wright Field.
The USAAF Air Intelligence organisation was built up from a very small nucleus in the days immediately before and after the entry of the United States into the Second World War on 7 December 1941. From small beginnings, and building on the experience of the Royal Air Force which had been studied in detail by USAAC personnel, a complex organisation developed, centred at Wright Field, Ohio.

Wilbur Wright Field had been established in 1917 on 2,075 acres of land adjacent to the Mad River which included the 1910 Wright Brothers' Huffman Prairie Flying Field and that was leased to the Army by the Miami Conservancy District. Logistics support to Wilbur Wright Field was by the adjacent Fairfield Aviation General Supply Depot established in January 1918 which also supplied three other Midwest Signal Corps aviation schools.

A Signal Corps Aviation School began in June 1917 for providing combat pilots to the Western Front in France, and the field housed an aviation mechanic's school and an armourer's school. On 19 June 1918, Lt. Frank Stuart Patterson at the airfield was testing machine gun/propeller synchronisation when a tie rod failure broke the wings off his Airco DH.4M while diving from 15,000 feet. Also in 1918, McCook Field near Dayton between Keowee Street and the Great Miami River began using space and

Wright Field OH in 1917.

mechanics at Wilbur Wright Field. Following World War One, the training school at Wilbur Wright Field was discontinued.

In 1924, the Dayton community purchased 4,500 acres, the portion of Fairfield Air Depot leased in 1917 for Wilbur Wright Field along with an additional 750 acres in Montgomery County to the southwest, now part of Riverside. The combined area was named Wright Field to honour both Wright Brothers. A new installation with permanent brick facilities was constructed on the new ground to replace McCook Field and was dedicated 12 October 1927. By November 1930, the facility at Wright Field had aircraft fitted as flying laboratories.

The Air Intelligence organisation called on the existing laboratory expertise at Wright Field in the same way that the RAE at Farnborough in the UK obtained the back-up of many specialists in a multitude of departments. The Technical Intelligence organisation was initially set up under the Engineering Branch of the USAAC Materiel Division - an Air Corps organisation which was responsible both for experimental engineering and for the Air Corps equipment procurement functions; it had its Headquarters at Wright Field until 1939. In October 1939 the Chief of the Materiel Division moved to Washington, DC, to be located with the Chief of the Air Corps, with his Assistant Chief remaining at Wright Field in charge of day-to-day Division activities.

From 1939 a considerable expansion and re-organisation of the Air Corps took place. On 15 March 1941 a Maintenance Command was set up to take over the Materiel Division's procurement activities and also the eight main Air Corps Depots.

This initially had its headquarters at Patterson Field. In October this organisation was renamed Air Service Command, to better reflect the fact

The 'new' Wright Field runways and associated taxiways under construction on 1 May 1944.

Wright Field in 1946.

that its logistics role included procurement as well as maintenance. Meanwhile, the US Army Air Corps, effective from 20 June 1941, became the United States Army Air Forces. The former Air Corps Chief, Major-General Henry H. 'Hap' Arnold, then became the Chief of the Army Air Forces and was given responsibility for establishing policies and plans for all Army aviation activities.

On 9 March 1942 the Materiel Division, which remained in charge of Wright Field experimental activities, became Materiel Command. At the same time the Wright Field experimental establishment was redesignated the Materiel Center.

Finally, on 1 April 1943, the functions of the former Materiel Division came together once more when Air Service Command and Materiel Command were merged into Air Technical Service Command (ATSC) with its headquarters at Patterson Field.

When the Materiel Division was up-graded to Command status, it became responsible for two major Divisions, Engineering and Production. The Engineering Division was responsible for many hundreds of research and development projects spread across its various Laboratories, varying from windtunnel tests of new aircraft configurations to the development of materials, aircraft ancilliary equipment and airfield equipment. Within the Laboratories were various Branches and Sections -for example the Aircraft Laboratory had a Glider Branch which carried out trials of troop-carrying gliders and their equipment.

Foreign Equipment was at first evaluated by the Foreign Equipment Unit of the Technical Staff within the Engineering Division.

Later, after the formation of the Technical Data Laboratory in December 1942, came the Evaluation Branch of the Technical Data Laboratory, giving rise to the 'EB-' series of numbers. The Foreign Equipment Unit then became part of the Evaluation Branch, which was in turn one of the six branches of the Technical Data Laboratory. The Technical Data Laboratory's tasks included '...*Procurement, evaluation and dissemination of technical information on foreign aircraft and aeronautical equipment, particularly enemy aircraft and equipment, of technical or military interest to the Engineering Division and American manufacturers producing AAF materiel*'.

Later the Evaluation Branch identities were changed to 'FE-' for 'Foreign

Equipment', presaging the formation of a Foreign Equipment Branch in April 1945 as one of Evaluation Branch's successor organisations - the Flight Data Branch being the other. Finally, in July 1945, another re-organisation saw the Technical Data Laboratory became part of 'T-2 Intelligence' within the organisational structure, with other 'T' designators indicating Personnel, Engineering, Administration and other functions. By November 1945, 'T-2' was transferred to a new Collection Division, under the command of Harold Watson, charged with the collection and assessment of equipment and technical information recovered from Germany, Japan and other sources. This Division remained in being until its task was completed in April 1947.

Although Wright Field was the centre for operations, the size of the Materiel Center and its successors greatly exceeded the capacity of the base. Because of this, several nearby bases were used to cater for the overflow. These included Patterson Field, Clinton County Army Air Field and Dayton Army Airfield (now Dayton International Airport). At its peak, the centre employed 50,000 civilian and military personnel. In 1943, personnel from Wright Field set up a further subsidiary unit at Rogers Lake, Muroc, California, known as the Materiel Command Flight Test Base, Muroc. This site later became known as Edwards Air Force Base.

Patterson Field

Patterson Field had been named after Lt Frank Stuart Patterson who had been killed on 19 June 1918 while testing a Airco DH.4M. It was designated on 6 July 1931 as the area of Wright Field east of Huffman Dam (including Fairfield Air Depot, Huffman Prairie, and Wright Field's airfield). Patterson Field became the location of the Materiel Division of the Air Corps and a key logistics centre and in 1935, quarters were built at Patterson Field, which in 1939 still *'...was without runways...heavier aircraft met difficulty in landing in*

Patterson Field in June 1941, just before the USA entered the Second World War.

One item captured from the Germans but not sent back to the USA for evaluation and exploitation was this decoy aircraft.

inclement weather'. The area's Army Air Fields had employment increase from approximately 3,700 in December 1939 to over 50,000 at the war's peak. Wright Field grew from approximately 30 buildings to a 2,064 acre facility with some 300 buildings and the Air Corps' first modern paved runways. The original part of the field became saturated with office and laboratory buildings and test facilities. The Hilltop area was acquired from private landowners in 1943–1944 to provide troop housing and services.

The portion of Patterson Field from Huffman Dam through the Brick Quarters at the south end of Patterson Field along Route 4 was administratively reassigned from Patterson Field to Wright Field. To avoid confusing the two areas of Wright Field, the south end of the former Patterson Field portion was designated 'Area A', the original Wright Field became 'Area B', and the north end of Patterson Field, including the flying field, 'Area C'.

In January 1948, Wright Field and Patterson Field were amalgamated to form Wright-Patterson AFB. The Collection Division was responsible for the operation of Freeman Field, where its Technical Operations Section looked after Restoration, Maintenance and Flight Research Branches.

The first Axis aircraft to be flown at Wright Field was Messerschmitt Bf.109E AE479, an example previously flown by the Luftwaffe, and then by the French Armée de l'Air and the RAF. This aircraft was shipped to the USA, arriving at Wright Field on 14 May 1942. Subsequent Axis aircraft were identified by the EB-, FE- or T2- numbers, but AE479 continued to be identified by its RAF serial number.

Later Axis aircraft were usually shipped in from the Middle East and Italy, but there are records of two Junkers Ju.88s being flown from the Middle East to the USA via the South Atlantic route.

In 1944, Col Harold E. Watson was attached to the USAAF in Europe and was responsible for the dispatch of further aircraft captured by US forces on the continent of Europe, culminating in the post-surrender shipment aboard HMS *Reaper* in July 1945. Aircraft were painted in US national markings at their point of surrender and flown to a suitable port for onward shipment to the USA, or, in exceptional cases, air-ferried or air-freighted to the US.

After the end of the war in Europe, Freeman Field, Seymour, Indiana, was transferred to the control of Wright Field for use as a subsidiary centre where many of the prizes received from Europe were re-assembled. The

Freeman Field, Indianna, as seen just after the war.

aircraft were flown from Freeman Field to Wright Field for evaluation testing to be carried out, since Freeman Field did not have equipment to support all flight tests under properly-controlled conditions. At this time, Freeman Field was called the Foreign Aircraft Evaluation Center. With the end of the war against Japan, the priority for assessment of German aircraft disappeared, since its main thrust was to ensure that any developments which might have been transferred from Germany to Japan were fully understood before they were met in combat during an invasion of the Japanese home islands.

Freeman Field
Freeman Army Airfield was named under War Department General Order Number 10, dated 3 March 1943, in honor of Captain Richard S. Freeman. A native of Indiana and 1930 graduate of West Point, he was awarded the Distinguished Flying Cross, and awarded the Mackay Trophy, and was one of the pioneers of the Army Air Mail Service. Captain Freeman was killed on 6 February 1941 in the crash of a B-17 Flying Fortress (B-17B 38-216) near Lovelock, Nevada while en route to Wright Field, Ohio. The aircraft was equipped with the top secret Norden bombsight and legend has it that sabotage was suspected but never was proven.

Initial surveys of the area were made in April 1942 and the present site of Freeman Municipal Airport was selected for construction. The selected site was announced on 3 April 1942. Army Air Force officials met with local landowners to obtain rights to a single tract of 2,500 acres for the main airfield and support base, along with five additional tracts for auxiliary landing fields near Walesboro, Grammer, St. Thomas, Kentucky, Zenas and Valonia, Indiana. Of the five auxiliaries, Walesboro and St. Anne were to have concrete runways.

The first construction for the new airfield began in late June 1942 with construction proceeding throughout the summer. It included more than one hundred buildings, all intended to be temporary. Station buildings and streets were also constructed, the buildings consisting primarily of wood, tar paper, and non-masonry siding. The use of concrete and steel was limited because of the critical need elsewhere. Most buildings were hot and dusty in the summer and very cold in the winter. Water, sewer and electrical services were also constructed. The airfield consisted of runways in a 'star' layout of four 5,500 x 150' runways laid out in an north/south, northeast/southwest, east/west and a northwest/southeast direction. An extra-large parking ramp was constructed to accommodate large numbers of training aircraft, several hangars, a control tower and other auxiliary support aircraft buildings.

The airfield was placed under the jurisdiction of the 33rd Twin Engine Flying Training Group, Army Air Forces Training Command. The 447th Base Headquarters and Air Base Squadron was activated on 2 October 1942, and the airfield was activated on 1 December 1942, and the first troops began arriving on 8 December 1942.

The mission of Freeman AAF was a twin-engine advanced aircraft training school. Most of the initial staffing cadre of the faculty was drawn from Craig Army Airfield, near Selma, Alabama. Five training squadrons, the 466th, 467th, 1078th, 1079th 1080th Twin-Engine Pilot were established at Freeman Field, and a total of 250 Beechcraft AT-10 Wichita trainers had arrived by the end of February 1943. The first flying cadets, who had just graduated from AAFTC advanced single-engine schools arrived on 2 March were formed as class 43-D. Night training commenced on 5 April. The first class graduated on 29 April and went on to fly multi-engine aircraft such as the B-24 Liberator, B-17 Flying Fortress, B-29 Superfortress, and various other medium bombers and transport aircraft. Twin-engine training continued with a total of 19 classes of students being graduated from Freeman Field. The last graduates were in May 1944 (Class 44-K); 4,245 cadets in total.

Twin-engine training ended in May 1944 and AAFTC initiated helicopter training at Freeman Field in June 1944. Freeman was the first helicopter base in the AAF. The first instructor pilots arrived on 30 June and preparations for the helicopter training were made in great secrecy, as in 1944 very few people had seen one and the technology was new and revolutionary. The group assigned to coordinate their arrival was known as 'Section B-O". A total of six Sikorsky R-4A helicopters were assigned for training, flown directly to Freeman from the Sikorsky plant at Bridgeport, Connecticut. This was the longest long-distance flight of helicopters at the time.

The first helicopter class began training in July, graduating on 13 August. The training programme continued throughout the balance of 1944, the last class (44-K) graduating on 1 February 1945. In January 1945, AAFTC moved the training to Chanute Field, Illinois, so it could consolidate the flying training operation with helicopter mechanic training.

With the end of helicopter training, Freeman Field's training mission was closed down and the facility was to be transferred as excess to Air Technical Service Command effective 1 March 1945.

When the 477th moved to Kentucky, on 2 May 1945, Freeman Field was placed on Standby Status, with jurisdiction of the facility being transferred to Air Technical Service

The sign at the entrance to Freeman Field a Fw.190 wing!

Command on 15 May.

In 1945 the enemy aircraft shipped to the United States were divided between the Navy and the Army Air Forces. General Hap Arnold ordered the preservation of one of every type of aircraft used by the enemy forces.

Initially, the Air Force brought their aircraft to Wright Field, and when the field could no longer handle additional aircraft, many were sent to Freeman Field to the foreign technology evaluation centre established there. Most of the foreign airplanes were German, but there were also Japanese, Italian and British machines present. Nowhere in the United States would there be such large numbers of foreign aircraft, many of which were rare and incredibly advanced for their time. In addition, there were warehouses full of Luftwaffe equipment. Forty-seven personnel were engaged in the identification, inspection and warehousing of captured foreign equipment. Freeman Field was also charged with the mission to receive and catalogue United States equipment for display at the present and for the future AAF museum.

The evaluation centre was the last United States Air Force operation at Freeman Field. By the middle of 1946, the programme was winding down and efforts began to dispose of the surplus captured equipment. Some of the Axis aircraft dispersed to Wright Field, the larger aircraft were sent to Davis-Monthan Field, Arizona, and the fighter aircraft sent to the 803 Special Depot, Park Ridge, Ill. (now Chicago's O'Hare Airport), under the control of ATSC's Office of Intelligence.

Not all of the captured aircraft assigned to Freeman were transferred. Some which were left at the field were destroyed or buried. Examples of aircraft that have no record of leaving Freeman Field are a Dornier Do.335 experimental interceptor; a Heinkel He.219 radar-equipped night fighter; an Arado Ar.234 twin-engined jet bomber, two Messerschmitt Me.163 rocket-powered interceptors, two Focke-Wulf Fw.190 interceptors and a Junkers Ju.88 two-engine multi-role aircraft.

A list of available Axis aircraft and equipment was published and the listed items released for study by approved US industrial organisations. The survivors were scrapped or moved to the Smithsonian Institution store at Silver Hill, Maryland, when the Park Ridge store was required for other purposes at the time of the Korean War.

Chapter Seven

'Operation Lusty'... and other 'Missions'

Technological change during World War Two had proceeded at an almost frightening pace. Developments in aircraft design, propulsion, weapons, and electronics all contributed vitally to the outcome of events in the global conflict. Although the British had undertaken formal technical liason with the United States of America from the first days of 'lend-lease' back in 1941 American involvement remained, at most, little more than passively documenting what crossed the Atlantic westwards.

There were, however, a number of scientists, largely civilians, to initially invent and design military equipment and then drive these developments forward to turn the tide of the war. After that came the might of production.

Among the scientists and thinkers was Hungarian aerodynamicist Dr. Theodore von Kármán. Since his arrival in the United States from Europe, having obtained Guggenheim funding and hoping to avoid rising nationalism and Nazism, he had become acquainted with several high-ranking Army Air Force officers, including Henry 'Hap' Arnold.

Since their first meeting at the California Institute of Technology (CalTech) in the early 1930s, Arnold had witnessed the professor's skilled use of mathematical equations to solve complex aerodynamic problems. Arnold's trust in Kármán grew as the CalTech programme continued to tackle the most difficult projects without hesitation.

During the early part of 1943, the Experimental Engineering Division of the United States Army Air Forces Material Command forwarded to von Kármán reports from British intelligence sources describing German rockets capable of reaching more than 100 miles . In a letter dated 2 August 1943 von Kármán provided the Army with his analysis of and comments on the German programme.

Only after D-Day and the realization of several key elements in wartime operations did Arnold believe that Allied victory in Europe was a foregone conclusion. The air war had become a deadly routine and was becoming a mere numbers game - growing Allied air strength versus dwindling Axis air capability.

By now General Arnold had decided that the Army Air Force was in a position to capitalise on the many technological developments. Following the shipment of several tons of captured German material back from France by US Intelligence not long after the start of Operation Overlord, Arnold realized that the United States and its Allies by no means led the world in military aeronautical development. He used his influence with Kármán at a super-secret meeting on the runway at La Guardia Airport, New York, convincing him to head a task force of scientists who would evaluate captured German aeronautical data and laboratories for the Army Air Force.

Kármán – who was recovering from recent abdominal surgery - was transported by Army Staff car to the end of the runway at

Dr. Theodore von Kármán (b May 11, 1881 – d May 6, 1963)

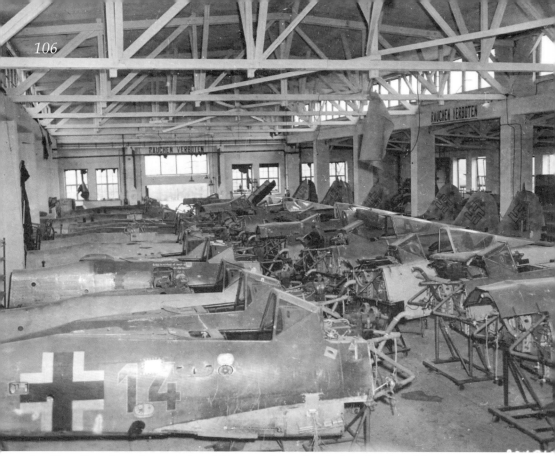

Focke-Wulf Fw.190 fuselages under construction at Kolleda, Germany in 1945. *[Hugh Jampton Collection]*

La Guardia where he met Arnold's recently arrived aircraft. Arnold dismissed Kármán's driver and then discussed his plans for Kármán and his desires for the exploitation project. Supposedly General Arnold spoke of his concerns for the future of American air power and wondered how jet propulsion, radar, rockets and other gizmos might affect the future. In response to the question as to what he wanted Kármán to do, Arnold told him that he wanted Kármán to go to the Pentagon, gather a group of scientists together and work out a blueprint for air research for the next fifty years.

To accomplish his mission, Kármán officially became an AAF consultant on scientific matters on 23 October 1944. His first, unofficial AAF report was organisational in nature, naming as his deputy Dr. Hugh L. Dryden, long-time head of the National Bureau of Standards. November 1944 saw endless conferences and establishment of '...relations with the various agencies in the labyrinth of military and scientific aviation'. Arnold drafted official, written instructions on 7 November, solidifying the LaGuardia Agreement, a four-page letter that set the boundaries for the report of Kármán's group. They were not very restrictive: *"Except perhaps to review current techniques and research trends, I am asking you and your associates to divorce yourselves from the present war in order to investigate all the possibilities and desirabilities for postwar and future war's development as respects the AAF. Upon completion of your studies, please then give me a report or guide for recommended future AAF research and development programs."*

Initially, Kármán's group was called the AAF Consulting Board for Future Research, but apparently AAFCBFR proved too long an acronym, even for the Army. Redesignated the Scientific Advisory Group (SAG) on 1

December 1944, it reported directly to General Arnold.

In 1944 intelligence experts at Wright Field had developed lists of advanced aviation equipment they wanted to examine. The U.S. Army Air Forces Intelligence Service sent teams to Europe hard on the heels of the invading armies to gain access to enemy aircraft, technical and scientific reports, research facilities, and weapons for study in the United States. The Air Technical Intelligence (ATI) teams, trained at the Technical Intelligence School at Wright Field, Ohio, collected enemy equipment to learn about Germany's technical developments. The ATI teams competed with 32 allied technical intelligence groups to gain information and equipment recovered from crash sites.

Enter the Navy!
In early 1944, it had become increasingly apparent that a centralised organisation was needed for the collection and dissemination of naval technical intelligence information. Inter-service rivalry made sure that the Navy were not going to miss out! There were numerous instances of the failure of such information to reach those who could benefit from it. Furthermore, with the intensification of the war in the Pacific, the fleet had an increasingly urgent need for timely technical information that had been processed into a standardised format.

Accordingly, Rear Admiral Roscoe E. Schuirmann, Director of Naval Intelligence, issued a memorandum on 3 October 1944 establishing the Technical Intelligence Center (TIC) within the Publications and Dissemination Branch and designating the centre OP-16-PT. The memo stated the duties of TIC to be as follows:
    a. Establish and maintain central technical intelligence files of all information relative to foreign warships, merchant ships, and naval and military equipment for the use of all service activities;
    b. Expedite and insure adequate routing and interchange of reports and information on these subjects;
    c. Maintain a panel in which representatives of all interested technical bureaus and activities will participate for the purpose of determining requirements of the activities either from incoming material or from the central files ;
    d. Collaborate with technical bureaus, through their representatives, in the preparation of technical intelligence articles on foreign naval equipment, on warships and on merchant ships, for dissemination through a common medium.

The Section served as the nucleus around which OP-16-PT was formed. One officer and one yeoman were also added from the Special Activities Branch (OP-l 6-Z) to control captured enemy equipment.

A Captured Enemy Equipment (CEE) programme, sponsored by the Bureau of Ordnance, assigned field personnel to both Europe and the Pacific, supplied them with cameras and CEE documents, and issued orders to photograph, identify, and serialise every piece of equipment sent back to the U.S. for exploitation. The field teams had the capability to write preliminary reports on CEE items and disseminate them rapidly to area units when appropriate.

Lt.Cdr C. H. Watson, USNR, was the first acting head of ONI's Technical Intelligence Center. The centre became involved in the filing and translation of documents of naval interest that were picked up by the Naval Technical Mission in Europe (NAVTECHMISEU) as elements of Europe were liberated. TIC functioned as a library and clearing house for the control of the unevaluated documents. A Control Section was established, and later an Estimates Section was formed, but, initially, no formal intelligence studies of the captured documents were undertaken.

The 'reading panel' system was adopted to help TIC personnel keep in close touch with representatives of agencies interested in the technical material available at the centre. Representatives visited the panel several times a week from the Office of the Chief of Naval Operations, the Office of Research and Inventions, all the Navy technical bureaus, the British Admiralty Delegation, the Joint Electronics Intelligence Agency, the Marine Corps, and the Military Intelligence Service of the War Department representing the Army. Dissemination was also made to the Naval War College, the Navy's General Board, the Ship Characteristics Board, and the Joint Army-Navy Experimental and Testing Board.

The US Navy made sure it was able to exploit German and Japanese Technical Developments. The Readiness Division of Commander Naval Forces, Europe (COMNAVEU), a unit that had performed well in the field of technical intelligence in England, prepared extensive plans for the exploitation of the vast sources of German technical information of interest to the Navy. Capt. Henry A. Schade was sent to Europe to investigate the best means of exploiting technical data about the German Navy. In May 1944, the War Department had proposed a joint Army/Navy project known as the 'Alsos Mission', for technical intelligence work in Europe. (Alsos, the Greek word for tree, was a play on words derived from the organization's having been established at the instigation of Army Maj-Gen. Leslie Groves, head of' the Manhattan Project.) The mission's primary purpose was to acquire the leading European nuclear scientists and data on the German atomic bomb project; the mission's other scientific data-gathering work was,

Raunchen Verboten! This Me.262 sitting in the hangar on oil drums is certain not smoking! Aircraft were discovered in this condition all over Germany. (USAAF)

in effect, a cover for its principal mission.

In August 1944, Capt. Schade was assigned as head of the Navy Section of the Alsos Mission, to report to COMNAVEU and to be under COMNAVEU administrative control. Alsos naval members were to represent COMNAVEU Readiness Division on the continent, and Commander Naval Forces, France provided assistance in personnel, billeting, and office space.

On 4 December 1944, the Secretary of the Navy approved the establishment of the U.S. Naval Technical Mission in Europe. Its mission was to exploit German science and technology for the benefit of the Navy Departments technical bureaus and the Coordinator of Research and Development. The mission's tasks were to coordinate all U.S. Navy activities on the continent of Europe that were exploiting German scientific and technical intelligence and to form a pool of technically qualified personnel under Navy control to operate as field teams, either independently or with Combined Intelligence Objectives Subcommittee teams, Technical Industrial Intelligence Committee teams, Alsos teams, or U.S. Army or British teams exploiting targets of naval interest.

The naval Alsos group that had been established to help in the search for information on, and personnel involved in, Germany's nuclear research served as the nucleus of the personnel pool. The senior Navy representative on the Alsos mission was designated by Commander in Chief, U.S. Fleet to be Chief NAVTECHMISEU The Navy technical bureaus and the coordinator of research and development provided technical officers, civilian technicians, and the necessary administrative personnel to staff NAVTECHMISEU. An office for the representative of NAVTECHMISEU was established in ONI (OP-16-R) to keep Chief NAVTECHMISEU continuously informed as to the plans and activities of the Technical Intelligence Committee and the technical missions of the War Department.

Chief NAVTECHMISEU was to report directly to COMNAVEU, and to the senior U.S. naval authority in the areas being exploited. He was authorized and directed to:

(a) travel and order travel, anywhere in Europe;
(b) obtain and expend funds as necessary in procuring technical intelligence;
(c) obtain necessary assistance from U.S. naval authorities in Europe;
(d) obtain assistance from U.S. Army authorities in Europe, using Alsos Mission channels wherever possible;
(e) forward Information Reports (IRS) direct to the Director of Naval Intelligence, with copies to appropriate Navy Department offices and to U.S. activities in Europe, and to communicate directly with the Navy Department regarding the intelligence operations of the missions;
(f) ship material to the United States of special interest to the Navy Department;
(g) return to the United States for consultation when necessary.

NAVTECHMISEU was activated on 20 January 1945. Commander Henry A. Schade was the first chief of the mission and was a direct representative of Commander in Chief, U.S. Fleet, reporting to Commander Naval Forces, Europe, with the designator Commander Task Force (CTF) 128. NAVTECHMISEU absorbed most of the officers from COMNAVEU's Forward Intelligence Unit, Task Group (TG) 125.8, and they became the Intelligence Section of the mission. Civilian technical specialists were provided by Navy contractors. One such civilian was the aviator Charles A. Lindbergh. The administrative headquarters for TG 125.8 was established in Paris, with forward headquarters located variously at Bad Schwabach (mid-April), Heidelberg (late April), Bremen (late May) and Munich (mid-July).

The personnel of the NAVTECHMISEU Intelligence Section (six officers and two enlisted) had been engaged in intelligence collection work on the continent since D-Day and were the most experienced naval field intelligence officers and men in the European theater. Their language qualifications, previous experience as interrogators of German prisoners of war, and familiarity with US Army field procedures were their principal assets.

At its peak, the Intelligence Section had thirty-eight officers and two enlisted personnel. The additional officers were recruited from CTF 124, the Special Activities Branch, and other naval activities, including the Bureau of Personnel.

Some interpreters were assigned on a semi-permanent basis to other NAVTECHMISEU sections, and about half were retained in an interpreter pool. Those assigned to a specific section made trips with officers of that section and later assisted in report writing and translating pertinent German documents.

With the AAF
In January 1945, Kármán's handpicked, scientific team of 'thirty-one giant brains' met in Washington to begin the task Arnold had given them. Kármán met internal resistance to a few of his choices for the group — for example, against an Englishman, Sir William Hawthorne. Colonel Frederick E. 'Fritz' Glantzberg, Kármán's military assistant, voiced his objection to having any 'foreigners' in the group. Kármán reminded the colonel that Arnold wanted the best people, regardless of their origins. Glantzberg eventually relented, somewhat grudgingly conceding that '...*the British are, after all, our Allies.*'

Sir William Rede Hawthorne CBE FRS FREng FIMECHE FRAeS, (b. 22

Almost certainly the same Me.262 that appears on the previous page. From the state of the engine cowling it appears that the aircraft had been dropped before being propped up on 50 gallon oil drums. *(USAAF)*

The hulks of Me.262 prototypes V9 and V10 at Lager-Lechfeld in June 1945.

May 1913 – d. 16 September 2011) was a British professor of engineering who worked on the development of the jet engine.

He was educated at Westminster School, London, then read mathematics and engineering at Trinity College, Cambridge, graduating in 1934 with a double first. He spent two years as a graduate apprentice with Babcock and Wilcox Ltd, then went to the Massachusetts Institute of Technology (MIT) in Cambridge, MA, where his research on laminar and turbulent flames earned him a ScD two years later.

After MIT, he returned to Babcock and Wilcox. In 1940, he joined the Royal Aircraft Establishment at Farnborough. He was seconded from there to Power Jets Ltd at Lutterworth, where he worked with Frank Whittle on combustion chamber development for the jet engine. Building on his work on the mixing of fuel and air in flames at MIT, he derived the mixture for fast combustion; the chambers produced by his team were used in the first British jet aircraft.

In 1941, he returned to Farnborough as head of the newly formed Gas Turbine Division and in 1944 he was sent for a time to Washington to work with the British Air Commission. In 1945, he became Deputy Director of Engine Research in the British Ministry of Supply before returning to America a year later as an Associate Professor of Engineering at MIT. He was appointed George Westinghouse Professor of Mechanical Engineering there at the age of 35, and in 1951 returned to Cambridge, UK as the first Hopkinson and Imperial Chemical Industries Professor of Applied Thermodynamics.

Kármán also insisted upon adding a naval officer to his team, William Bollay (a former CalTech student). When the colonel insisted that the professor had gone too far, Kármán responded with the simple question, *"But Colonel, the Navy are surely our Allies too?"* After considering this for a moment, Glantzberg finally agreed that they were *'...Not as close as the British, but a damn sight closer than the Russians."*

SAG meetings held during the first weeks in February, March, and April accomplished the basic research and finalized the general format for the report. Kármán emphasized that these spring meetings had a threefold purpose: the SAG would search for ways to secure *'...scientific insight in a standing Air Force'*, it would ensure the continued interest of American scientists in the future of the Air Force; and the group would educate the American public in the necessity of maintaining a strong Air Force.

After the scientists arrived in Paris on 1 May 1945, one team-member, H Guyford Stever, observed the critical nature of the timing of the Allied advance. In 1942 Stever had begun serving the military as a civilian scientific liaison officer based in London until the end of the war. After the start of Operation Overlord he was sent to France several times to study German technology. He noted that although local looting often presented a problem, it was the advancing Russians that presented the real concern. Significantly, Stever noted that '...*until the von Kármán mission we scientists had to piece the enemy's facts together. Now we had the advantage of actually talking to the German scientists and engineers, seeing their laboratories, and hearing them describe their total programs*'. Dr Dryden concurred: '... *I think we found out more about what had been going on in a few days conversations with some of those key German leaders than all the running around and digging for drawings and models.*'

As the Allies advanced into Europe during the spring of 1945, Kármán's team, close on the heels of the advancing wave, scoured German laboratories.

As the war concluded, the various intelligence teams, including the ATI, shifted from tactical intelligence to post hostilities investigations. What was termed 'Exploitation Intelligence' - described as taking full advantage of success in military operations, following up initial gains, and making permanent the temporary effects already achieved and taking full advantage of any information that has come to hand for tactical, operational, or strategic purposes. - increased dramatically.

Exploitation Intelligence was also defined as the process by which captured enemy personnel and materiel are exploited for intelligence purposes as part of the all-source effort to provide 'decision advantage' to decision-makers at all levels.

Dr Millikan reports back
One of the scientists that travelled to Europe was Dr Clark Blanchard Millikan (*b* 23 August 1903 - *d* 2 January 1966), a distinguished professor of aeronautics at the California Institute of Technology (Caltech), and a founding member of the National Academy of Engineering. He attended the University of Chicago Laboratory Schools, graduated from Yale College in 1924, then earned his PhD in physics and mathematics at CalTech in 1928 under Professor Harry Bateman. He became a professor upon receiving his degree, full professor of aeronautics in 1940 and Director of the Guggenheim Aeronautical Laboratory in 1949.

His first major engineering work began with the construction of large wind tunnels, particularly the Southern California Cooperative Wind Tunnel in Pasadena, which was shared by five major aircraft companies. CalTech wind tunnels were subsequently used during the design phase of more than 600 types of aircraft and missiles.

He was active in the formation of the Jet Propulsion Laboratory during the war, and served as chairman of CalTech's Jet Propulsion Laboratory committee from 1949 onwards.

Upon his return from Europe, Millikan discussed his findings with other members of the Board of Aeronautics, and the transcripts of this meeting make facinating reading, for they show just how far ahead in his opinion the Germans were. ' *I thought you would be interested in the elaborate, in fact extremely elaborate research setup which has been in effect in Germany for the past ten years or so. There has been very numerous research establishments upon quite a large scale under the auspices of the aircraft ministry which is called, abbreviated for short RLM, Reichs Luftfhart Ministorium. I thought I might list a few of the major ones and describe a few of the facilities that are in some of these establishments*

Doctors Theodore von Kármán, Clark B. Millikan and Arthur L. Klein at California Institute of Technology

so that you can see the magnitude and scale on which the Germans have this aeronautical research set-up. The chief laboratory which I would presume would be called the DVL was located in Berlin and was a sort of central laboratory to the DLM. It might correspond to a sort of combination of our NACA Laboratory and Wright Field and the Navy's establishments sort of rolled into one, for aeronautical research. First I'll list just four or five major laboratories and then discuss them a little. The first of importance is probably the famous Herman Goering Institute at Volkenrode just outside Brunswick. Then there is the so-called AVA (Aerodynamische Versuches-Anstalt) at Gottingen which was sponsored by the Aircraft Ministry and adjacent actually on the same grounds, is the entirely independent laboratory under Professor Prandtl, the so-called Kaiser Wilhelm Institute which has been a separate institution for many years. It now appears as if it was more or less absorbed, but Prandtl insists that it is essentially an independent agency and has nothing to do with the AVA which is under Dr Detts direction. Then there's a very ambitious project, which was in the construction stage at Munich. Finally, the sailplane institute under Dr Georgii which was originally at Darmstadt and has been successfully evacuated and finally wound up in Ainring in the Austrian Tyrol.

'These are all essentially Aircraft Ministry establishments and as you will see, each is on a fairly large scale. In addition there are the establishments which are under the auspices of the War Department, and the SS had a few of them, there was quite a number of other agencies, the only one I'll mention is the Peenamunde establishment which developed the V2, or A4 rocket as the Germans call it and which was evacuated from Peenamünde down through Germany and finally wound up at Garmisch and Kochel down in the Tyrol and was independent from the Aircraft Ministry.

Dr. Ludwig Prandtl
(*b.*4 February 1875 –
*d.*15 August 1953)

'The DVL I know very little about and I think that very few people this side of the water know much about it because, being in Berlin, it was taken over by the Russians and as far as I know has not been visited by any of our technical people at all. We do know the names of the people who were there. Several of the good ones actually got out of the Russian Zone and are somewhere in the American Zone now, so it's possible that some of them may actually be brought to the States. In any case, it was a very large and very well equipped laboratory.

The Herman Goering Institute at Volkenrode is an enormous place. Apparently it was kept from the Allies knowledge almost completely although the Germans said the Russians did know something about it. It was built in a wood , in three woods as a matter of fact, and covers an enormous area, with the most elaborate camouflage I have ever seen. Even though the sides of the buildings aren't camouflaged, the woods are very thick on top, and on one of the tunnels they have three feet of sod all planted with a forest growing on top. I've seen the bombing of the reconnaissance photographs and apparently we never knew it was there at all. In any case it was never bombed, although it is right adjacent to an airport. The airport was bombed but the experimental establishment was untouched. It has employed several thousand people. There were seven or eight large wind tunnels up to twenty feet in diameter, the working section of the largest one was a little over twenty feet in diameter. There were four very good supersonic wind tunnels. There is a very elaborate and complete engine test installation off in a wood by itself, with perhaps half a dozen buildings for the engine division, completely separate from the aerodynamic and wind tunnel.

There is another division in a separate wood which deals only with armament problems, which has I suppose the most elaborate ballistic range anywhere in the world, underground tunnels that are very long, a good many thousand feet long with very elaborate ballistic recording apparatus in them, and with little wind tunnels blowing sideways across the ballistic range to give the effect of side winds and yaw on bullets. That's an enormous and elaborate place that I just did not get to because there was so much to see there. In addition to that, there is a rocket research establishment that is separate under Dr. Duseman and some miles away, with a large mathematic and theoretical section.

It's a really tremendous place and they told us that they had all the money that they could possibly spend, and were asked to come in with recommendation with more to spend which they could not use. They said they couldn't use their appropriation up even in the last couple of years. They were apparently very elaborately supported. The establishment at Göttingen does not have nearly the same physical equipment as the Bruinswick tunnels. It has one big twenty foot tunnel, but then a very large number of small special purpose experimental tunnels and other

An Bf.109 fighter in one of the large German wind tunnels.

facilities that are set up to investigate special problems.

There are three or four tunnels in which they are doing a lot of work on non-stationary motion. They are doing a lot of flutter work by oscillating heavy metal airfoils in water, and they get a very good reproduction of the normal aeroplane characteristics. They are apparently using that technique a great deal. They have a lot of supersonic tunnels, small supersonic tunnels, two or three of them a pretty good size. They have a cavitation tunnel used in conjunction with high speed research they are doing. They have a very complete installation for testing turbine and compressor blades and all sorts of thermodynamic and heat transfer problems are investigated, and probably the best collection of scientific brains that there are in Germany, if you take both the people under Bett's direction in the AVA and Prof. Pradtl's group at the Kaiser Wilhelm Institute. So, although it isn't on so grandiose a scale as Brunswick, it was certainly one of the very effective institutions in Germany.

Munich was the brainchild of Bäumke's I understand. Bäumke was the director of the Aircraft Ministry until the last year or so, and he felt that there should be in the south of Germany an elaborate installation set-up analogeous to the Herman Goering one in the north, and consequently they established this very elaborate one in Munich. It was only under construction at the time of the American occupation. They have one nine feet diameter wind tunnel which reached Mach One with about 12,000 horsepower. They have three or four supersonic wind tunnels that were under construction, one of them perhaps the most advanced supersonic tunnel in Germany, which actually a Navy Lieutenant and I discovered and tagged and that last I know it still had Navy tags on it but maybe the Army's taken them off by now.

In any case, it was sitting on the cars and has never been assembled. The compressor and all the equipment was furnished by Brown-Bavuria, Switzerland. So there is a continuous-operation supersonic wind tunnel that's just waiting there for someone to pick it up and ship off and assemble it somewhere else. It has never been used.

They also have very elaborate engine and rocket motor test installations planned. The buildings were half up. They had in the motor laboratory as I remember, ten test stands and in the rocket laboratory another ten test stands, so they could carry on ten simultaneous tests on either rocket motors or jet propulsion or gas-turbine types of motors.

The most fabulous of the developments was the largest wind tunnel, I suppose the most powerful in the world, which was being built under the auspices of the Munich Institute but down in the Tyrol just close to the Italian border, in a place called Oztal. It was known as the Oztal Tunnel, it has a working section diameter of 20 feet, reached a Mach number of 1 and was driven by 100,000 horsepower which was obtained from two 50,000 horsepower pelton wheels. They put in an enormous hydraulic project for 500,000 kilowatts, even put in a dam which was under construction and 400,000 of the kilowatts would go for electric power and about 100,000 were taken to drive this pelton wheel to drive the wind tunnel. That wind tunnel had been under construction for several years. It's usually called 70% complete although the actual steel structure is somewhat less than 70% complete, but apparently all the machinery was finished. The two pelton wheels are sitting there. The buildings are largely complete and would have been ready to operate within six months. It was equipped for handling complete jet powerplants up to at least 4,000 pounds of thrust with full power and running at a Mach number at least as close to 1 as you could run without having tunnel choke. It is a really gargantuan undertaking.

The only other one I want to mention is the Army one, which at the time was in Kochel having been evacuated from Peenemünde, which was primarily a ballistic research establishment. Practically all of the work done on the German V2 was done by the Peenemünde group. It was a very extraordinary group. There was no geniuses as far as I could discover among them but a large number of highly competent and qualified engineers and scientists. But as I say, none of them are outstanding, they turned out what to my mind was the greatest volume and the finest supersonic tunnel work of any place in the world. I think the material they turned out during the course of this V2 aerodynamic research is simply remarkable and we are fortunately going to have a complete record of it.

The Army had Dr Zwicky as a Technician and he got them to work at 08.00 in the morning and cracked the whip and made them all finish up the reports they were writing. They had buried most of their documents and he made them dig them up, put all the photographs back in. We're going to have 183 reports out of Kochel and a nice volume when Zwicky gets through with it, which is about now. That project was actually handled better than any of the establishments we had moved in on.

They had by the way two supersonic wind tunnels about fifteen inches square in cross-section which ran intermittently. They evacuate a great big sphere down to a very low pressure and then let the air rush through the tunnel into the sphere and they can run for about 20 seconds during which time they get their observations. They have the highest Mach numbers of any of the tunnels actually running in Germany going up to nearly four times the speed of sound, and have actually done a simply enormous amount of work on projectile shapes, thin small wings on projectiles.

He then described his discoveries in the field of aerodynamics, including the German discoveries of wing sweepback and wing high lift devices.

'Conventional flaps don't show up very well with high sweepback and the Germans have developed, especially for these, very thin pointed wings which are necessary at high speeds; it's a new kind of flap which is a nose flap with a little piece of the leading edge simply flipping down from the leading edge of the wing. [Today, this is called a leading edge slat]. Using it, the lift actually goes up quite high, and it's going to be something we're going to use a lot in the future. It's one of the fields where we have a lot to do and we've spent a lot of time and gotten a lot of effect in reading all the reports the Germans have written, because there are a lot of them on this whole sweepback technique.

Another one which to me is very interesting, which is a little time off, is the development of ram-jets, or jet powerplants without compressors. There were a lot of people working on that, but there were two developments which I think were outstanding; one was by the Folkewolke Co. a young fellow by the name of Pabst being responsible for it. He had first gotten a new conception of how a ram-jet works, which I think was a very fruitful conception. The ordinary calculations for ram-jets are usually done by calculating the momentum coming in, the momentum coming out and the thrust is the difference, the excess of the momentum going out over the momentum coming in and that has always been somewhat unsatisfactory to some of us and it was entirely unsatisfactory to Pabst. He said that you don't ever get a thrust concept when it is a result of some pressures acting over a surface and he was not satisfied just calculating these overall momentum exchanges. He said where does the pressure act, and where does the actual force exist on the ram-jet and what he discovered was a thing that one would think of if he thought hard enough and clear enough, immediately, is that the pressure, the suction, over the leading edge of the ram-jet. It is exactly like a NACA chord and the effect of the combustion inside it primarily to put a blocking effect in, which gives you a stagnation point on the inside of the chord with a very high velocity over the nose and he calculates the pressure distribution and gets the resultant forward force, the thrust and his ram-jets are very reasonable, sensible looking things.

They look very much like well designed NACA cowls, they're short and fat instead of those great long things that everybody else was working on for the ram-jets. That was one very important development, and another is that he [Pabst] became convinced that because of the internal shape of the ram-jet is so important, you could not have those terribly long things and therefore it was necessary that the combustion chamber be shortened, so he completely abandoned the idea of combustion of liquid fuels. Pabst uses only gaseous fuels and when he uses gasoline

A V-2 rocket about to launch.

he uses a separate little combustion chamber and vaporizes and preheats the gasoline before he feeds it into his combustion chamber as a gas. He gets 100% combustion in about 20 centimeters. These things are very remarkable and that has been tested in the high speed wind tunnels up to Mach 0.9. It's thought that Focke-Wulf was working on planes using the powerplant.

The other is a still more fascinating development, which is the high Mach number supersonic ram-jet that Oswatuch from Göttingen has been working on, and he said the essential problem here was to get a supersonic diffuser that will give you a large pressure recovery and he had actually developed a supersonic diffuser at a Mach number of 3 which gave him twice the impact pressure that you would get with an ordinary Pitot static tube stuck out there in the free stream. His design causes a series of oblique shock-waves to occur so that you get the velocity successively stepped down until just before the entrance to the actual closed diffuser the velocity was only slightly above the speed of sound and then had a normal shock wave with a relatively small entropy increase and got this very efficient supersonic diffuser.

In his calculations from the basis of his experimental data at Mach 3 he got an overall thermodynamic efficiency of the ram-jet of between 40-50% which is a fantastic figure. The compression ratio that he gets in a Mach number of 3 is 19 to 1, which makes the high thermodynamic efficiency not unreasonable. That's something that I think very definitely is going to be pushed very rapidly and we have his complete reports and designs and I think we can go right ahead and work on it.

From this original concept came 'Inlet cones' (sometimes called shock cones or inlet centerbodies) which became a component of some supersonic aircraft and missiles. They are primarily used on ramjets, such as the D-21 Tagboard and Lockheed X-7. Some turbojet aircraft including the Soviet Su-7, MiG-21, English Electric Lightning, and SR-71 also use an inlet cone. Another version of the same basic concept was the intake ramp, as used on such aircraft as the Concorde, XB-70 Valkyrie and F-15.

Dr Millikan then changed the topic to the developments they had discovered in the turbine field.

There are a number of developments that are not yet that well known by our powerplant people. One that was done at Göttingen was a ceramic heat exchanger for gas turbines whereby the heat of the exhaust gasses from the turbine is taken out and fed into the air coming from the compressor before the combustion chambers so that you get a regenerative turbine cycle. The weight of this installation is very light. The weight is just about 3/10 of a pound per horsepower which is fairly heavy for a jet engine but the thermodynamic efficiency is up around 40-50%. A 5000 kilowatt shaft power turbine has been developed to power the Me.264. The Me.264 was that long range bomber that they were talking about coming over and bombing New York with, and the turbine for that was to deliver 5000 kilowatts shaft power was to have the Ritz heat eschanger on it and the fuel consumption calculated on the basis of his thrust horsepower per horsepower hour because of the comparable figure of a 4-cycle Otto engine and propeller that gives is 0.48, which the fuel consumption is cut down to not very much over, well, it's about 2/3 of what our present fuel consumption is with this device.

Ritz is a first class scientist, but he is not as practical an engineer as the next man that I'll mention, Dr Schmidt. I think certain developments have progressed far enough and looks as if it has enough promise to be worth very very careful investigation by experts who are far better able to judge it's value than I am.

The other gas turbine development was done by Dr Schmidt who was the director of the engine laboratory at Brunswick at the Herman Goeing Institute. What he was trying to do was to get higher combustion temperature as in Germany they were very limited on alloys. Everywhere you went they were struggling to make the cheapest kind of materials work for them and they apparently had no alloys to make high temperature resistant steels, so his problems was to get high temperatures with the special alloys and he did this by a very ingenious system of water cooling. He drills three little radial holes in each of the blades of his turbines and water is forced through the hollow shafts and it gets into these little radial holes and the surface is hotter than the center of these little holes because of all the heat coming from combustion and so you get a slight difference in the density of the water on the edge which is heated and the water in the center which is not so heated and in the

In 1944 there was a demand in Germany for a supersonic wind tunnel. Therefore a group of German engineers began building a wind tunnel, at the city of Ötztal, in the former Austria. The 8-foot diameter wind tunnel, would be able to obtain wind speeds close to the speed of sound. Water from the Ötztal Ache was supplied through a tunnel and a water turbine drove a hydrodynamic wind tunnel at the foot of the mountain. It was intended that the water-driven turbine should be able to generate 75 MW, which was large enough to obtain the desired effect. 3500 forced labourers, including Russian prisoners of war, were put to the task of construction. They lived in miserable conditions in camps and many died under the inhuman conditions. The plan was that the wind tunnel should be completed in autumn 1945 and work continued until the end of April, 1945. All the secret plants that the Germans established in existing mines were code named after sea creatures. The facility at Ötztal - seen here under construction - was codenamed Zitteraal, or Electric Eel. In August 1945, all the equipment was dismantled and moved to Mondane in France.

tremendous centrifugal force field due to the rotation - it gets up to 2000g - you get a very intense circulation with the water coming out of the sides of these holes and down the middle and out near the other edge of the blade where you need the cooling the most, the conditions are such that with this tremendous pressure the water is actually about at the critical stage and he says that there was plenty of cooling to carry the heat away from the blades without any difficulty at all.

The water is then forced down into the middle and as it gets into the shaft the pressure is reduced and it is driven out of the other end of the shaft as steam where it drives a little auxiliary turbine and is cooled down and goes back into the cycle.

The MAN company which is one of the largest rotating machinery companies in Germany, has contracted and was building a 10,000 kilowatt stationary powerplant using these water-cooled blades in which they had guaranteed an overall thermal efficiency for the plant of 35%. They expected to get up close to 40, but it guaranteed 35. He was able to run his combustion chamber, his combustion gases, at 1200 degrees centigrade which I think turns out at 2300 Farenheit, with ordinary mild steel blades. As far as I know the highest temperatures we can reach are in the order of 1500 Farenheit, so he's pushed the temperatures up 500 or 600 degrees with this thing, but it seems to me that both developments are well worth investigating by our power plant people and may really cause come considerable revolution in our power plants.

I think perhaps I have one other item to talk about, which is research aeroplanes. There isn't very much to say about that except that both the Germans and the British have a very active program for developing aeroplanes to go up into this supersonic region.

The Germans had one designed by Lippich, which was under contract by Sieble, designed to reach a Mach number of 2 or 3 with rocket motors of the type that power the Me.163. That was a very carefully laid out design with a lot of very interesting features on it. Junkers was developing an improved 163 which I think they hoped to get up to the speed of sound, or a little through it. Messerschmitt was doing a very elaborate series of research aeroplanes in an attempt to push the 262 up higher. They weren't hoping to get through the speed of sound. They were perhaps closer to technical aeroplanes than these others, but it's felt that people are very much interested in and a lot of work was being done in Germany and I myself feel that a lot of work should be done here because I think the problem that we're running into as we get into these small supersonic velocities or high subsonic velocities are so difficult that we just won't solve them unless we get some aeroplanes up that actually fly at these speeds.

With the main speaking part of the meeting over, it was time for questions and answers, which immediately turned to practical things. Unfortunately the transcripts fail to reveal the identities of most of his questioners.

Question: What's going to happen Doctor to all this gear that is over there, wind tunnels and that sort of stuff, is there any plan made for the shipment?

Dr. Millikan: The question as to what's to happen to the wind tunnels is apparently still very much up in the air. I gather that a meeting is being held early next week which as far as I know is the first time that the various agencies interested have gotten around a table and there will be some table-banging as to who is getting what.

As far as I know plans for evacuation have gotten so far that everybody agrees that some German personnel are to be brought over here. Just what the mechanism or how they are to be treated or handled I don't believe anybody has faced yet or solved. It's a very difficult problem.

An unidentified speaker – from the phraseology almost certainly a Naval Officer - then explained what they knew: 'That is being handled by OP16, Naval Intelligence. We're going into that matter in order to bring it under proper control and expect that they will handle these people that are brought over and make them available to us as desired. Navy Intelligence Center OP16PT received all information

Another Bf.109 in one of the large-diameter German wind tunnels. This aircraft appears to have had it's tailplane removed for testing.

and captured enemy material and the scientists are considered no more or no less than captured enemy material and will be treated as such.'

Dr Millikan: *There is certainly a lot of material that is valuable, no question about it. We've had very active people both Navy, the Air Force and Ordnance, have pulled out a lot of stuff and put it in good safe American territory for further distribution.*

Someone else: *'Well that is what I had in mind. While all the meetings are being held in Washington, I think somebody over there ought to be trying to get a hold of some of this stuff'.*

Dr Millikan: *'They have been getting a lot of it. As far as know those Kochel wind tunnels hadn't been moved when I last heard. We sent down just when I left a team to work on this Munich wind tunnel. We got a hold of the fellow who did the designing, he was a former MIT professor who was there and was very cooperative with inventorying the whole thing and telling us just what parts belonged together out of this terrible mess. There was one installation that was very seriously damaged, but not by fighting, it was the displaced persons and mostly GIs. They came in and were billeted there and there was nobody to protect the place, and it's a mess. Everything is dumped together – it will take weeks to sort it all out. We recommended that they got 10 or 12 German engineers that were sitting around doing nothing who had actually constructed this thing and put them to work sorting it out.*

Question: *'Doctor. From your observations over there in regard to the larger pieces of equipment etc, which would be impractical to transport over here, what would your opinion be in regard to the possibility of our establishing a unit in order to operate and see the value which can be obtained, until we duplicate them in this country?*

Dr. Millikan: *The British are talking about doing that at Volkenrode. You see that is in British Territory. We had it, then the British moved in and took over and there is talk of them operating it as a British Establishment. I myself am inclined to think it a little dangerous. I think we just don't want such big research centers left in Germany and I think we ought to pull them out or blow them up.*

The British 'Mission'
Clearly, although well-hidden in the Millikan Report, there had been clashes - if not physical, then ideologic - between the British and Americans. The underlying sub-text of all of this is not fully revealed until one studies the work of a number of British teams who were attempting to discover the

same things at the same time.

'The Fedden Mission' was a British scientific team sent by the Ministry of Aircraft Production to Germany to gather technical intelligence about German aircraft and aero engines and was named after the Mission's leader, Roy Fedden. Sir Alfred Hubert Roy Fedden MBE FRAeS who was an engineer who designed most of Bristol Engine Company's successful aircraft engines.

Organised at the instruction of Stafford Cripps, then Minister of Aircraft Production, the Mission consisted of Fedden; Dr. W. J. Duncan, Professor of Aeronautics from University College of Hull (then seconded to the RAE); J. C. King of RAE's Structural and Mechanical Engineering Department; Flight Lieutenant A B P Beeton, RAF, of RAE's Engine Department and Bert Newport of Rotol, Ltd. They were assisted by W. J. Stern of the Allied Control Commission and Wing Commander V. Cross, RAF, the Mission's Liaison Officer to Supreme Headquarters Allied Expeditionary Force (as well as its translator); their two RAF Dakotas were flown by Flt Lt Reid, RAF, and Flt Lt Cheany, RAFVR.

Everywhere the Mission went, it encountered evidence of looting by Allied troops and German civilians, and sabotage by German factory workers. The German scientists and technicians were, in general, very co-operative with the British interviewers, with Fedden mentioning that they had 'considerable apprehension' about their fate and some expressed strong desire to emigrate to the U.S. or Canada. It seems that they had genuine concern about forced emigration to the Soviet Union, as many of the V-2 program scientists suffered.

The Mission left RAF Northolt on 12 June 1945, bound for Bückeburg, Germany, 25 miles southwest of Hannover. They were immediately sent on to another airfield and eventually reached Bad Eilsen, headquarters of Second Tactical Air Force. From there, they were driven to 21st Army Group headquarters at Bad Oeynhausen, where they spent their first night. Wing Commander Cross, by proposing one aircraft fly ahead to arrange accommodations, permits for interviews, travel permits for interview subjects, and so on, made an important contribution to the timely operation, and ultimate success, of the Mission. The Mission's access to German civilians was strictly limited by an Allied policy discouraging contact beyond what was strictly necessary.

Sir Alfred Hubert Roy Fedden MBE FRAeS (*b* 6 June 1885 – *d* 21 November 1973)

The Mission based their initial forays out of Braunschweig, in the British Occupation Zone, where they spent their first six days, moving later to Kassel, in the American Zone. For their more southerly objectives, they worked out of Freising, headquarters of Third U.S. Army Intelligence, some 20 miles from Hamburg, for another six days. Their third base of operations was SHAEF Headquarters, Frankfurt.

On their second full day in Germany, 14 June, they visited the Herman Goering Institute at Völkenrode where they remained through 15 June. Völkenrode was at the top of Fedden's list, because it was in the designated Russian Zone of Occupation, due to be handed over from British control in a matter of days. When the Mission arrived, the Institute had already been stripped by Combined Intelligence Objectives Subcommittee (CIOS) teams, and was (despite being in the British Zone) effectively in U.S. hands.

In Fedden's own words: *'We arrived at a particularly difficult time. Two days before we started, the re-orientation of the areas to be permanently occupied by the various allies had been announced and when we arrived a sort of general post was in full swing. The Russians were taking over large additional areas, the Americans were moving out of some places and the British from others and considerable confusion reigned. The advance of the Red Army meant that some important targets had to be*

struck off our schedule but we managed to visit 33 places.

At Völkenrode, the Mission examined the wind tunnels used to develop the swept wing, the forward-swept wing, the Messerschmitt Me.262, cowlings for the BMW 801 radial (used in the Focke-Wulf Fw.190), the pulsejet engine of the V 1, and other items, examples of the Rheinmetall-Börsig F25 and the Amerikabomber, as well as the Engine Department. While finding trials of photoelastic lacquers for stress tests, it proved impossible to interview any of the laboratory staff to learn more. Fedden considered the LFA Engine Department lacking in modern equipment, but a better altitude test bed than any in Britain; a better example still would be seen at BMW's Munich facility. The engine research included turbine and stator blade forms, blade cooling, blade construction (including hollow and ceramic types), piston cooling, and other matters. While at Völkenrode, the Americans stole equipment from under the noses of the British (including an interferometer), only to have the thefts denied when confronted by Britain at Potsdam. The Fedden Mission spent two days at Völkenrode, its sub-team returning ten days later.

The Mission left Braunschweig airfield on 15 June for Kassel, 80 miles away; next morning, one group went to Göttingen to interrogate Dr. Ludwig Prandtl and his team; Fedden called them *'...a first class team of experimental research workers'.* While in Göttingen, Mission members examined the wind tunnels on the campus, plus several used for fluid dynamics studies. They spent 17 June in Göttingen, also.

The same day, the second Mission group flew to Oschersleben, then drove by Jeep to the Junkers jet engine plant at Magdeburg. They had little time before it, and all the surrounding territory, was handed over to the Soviets. The Junkers works, used for manufacturing and overhaul of Jumo 004 jet engines, had been heavily bombed; while there were a number of engine test-beds, there was no sign of jet engine research having been done there. Otto Hartkopf, then acting works manager, conducted the Junkers factory tour, explaining all the jet engine drawings had already been removed. Hartkopf reported that over 5,000 jet engines had been produced there in all; at war's end, deliveries of 004s had reached 1,500 a month, but production, including several other plants, was expected to reach 5,000 per month. The Mission observed the construction methods of the 004.

The Mission also examined production of the BMW 003 jet engine at

Prior to the German surrender, this was really the only familiar view of a Me.262, with undercarriage down about to land - which was the only consistent time Allied fighters were able to catch and shoot down the German jets. *(Peter H T Green Collection)*

facilities in Eisenach and Stassfurt. On 18 June, it split into two teams again, one driving the Eisenach works. Fedden met with Dr. Bruno Bruckmann, head of BMW's jet engine research program in Berlin and strong supporter of jet engine production and use, as well as to drive propellers (turboprop engines); in 1942, Bruckman was made head of BMW piston engine programmes. At Eisenach, the Mission spoke to the facility's managing director, Dr. Schaaf, and Drs. Fattler and Stoffergen, learning that BMW employed 11,000 there in all, 4,500 in a camouflaged factory in the side of the hill, the rest in the town. Despite adding plants at Spandau, Nordhausen, and Prague, BMW never reached the production target of 5,000 to 6,000 003 engines a month.

At Eisenach, the Mission discovered the BMW 003A had incorporated a reusable liquid fueled rocket engine in the rear of the nacelle, the BMW 109-718, to act as an assisted take-off unit, or to provide acceleration in climb or flight - something that was akin to what the Americans postwar called 'mixed power'. Fedden called the production quality at Eisenach 'excellent'. Next day, the Mission examined a BMW facility near Stassfurt, set up in a former salt mine 1,300 feet underground, which was to have been used for machining jet engine parts, and possibly for assembly; Stoffergen said 1,700 machine tools had been installed, and some 2,000 workers had been employed. The Mission also found some information on the BMW 018 jet engine project, which was begun in 1940, but remained unfinished by war's end.

On 21 June, the Fedden Mission travelled from Eisenach to Klobermoor, location of the Heinkel-Hirth engine works. There, they examined copies of the Heinkel HeS 011 jet engine, said to be one of Germany's finest and most advanced turbine engines of the period. The Mission conducted extensive

Dornier Do.335s in the final assembly hangar at Oberpfaffenhofen, not long after the site was overrun by advancing American troops in 1945. [Hugh Jampton Collection]

German airfields became massive aircraft dumps, as seen here. *(Peter H T Green Collection)*

interviews with the managing director, Mr. Schaaf, and the senior planning engineer, Mr. Dorls, as well as Hartkopf, comparing piston and jet engine production; they compiled a table of comparative cost of materials, finding jets were between one half and two-thirds as costly, as well as being simpler and requiring lower-skill labour and less sophisticated tooling; in fact, most of the making of hollow turbine blades and sheet metal work on jets could be done by tooling used in making automobile body panels. While Fedden was critical of some of the German design decisions, the Mission estimated German jet engine production by mid-1946 would have been at a rate of at least 100,000 a year.

The team was told the overhaul cycle for the Jumo 004 was between thirty and fifty hours (and about fifty for the BMW 003), and approximately 300 Jumo 004 engines had been rebuilt, some more than once.

On 18 June, the Mission drove from Kothen to Dessau, home of Junkers Flugzeugwerke. Speaking to the technical director in charge of the 213's development, they learned there were many variants, but only three models: the 213A - the major production version the 213E (a high-altitude model), and the projected 213J (improved still more).

The Mission visited Nordhausen (the Mittelwerk) on 19 June, some by Dakota, some by Jeep. There, they first came in contact with concentration camps which revolted the Mission members. With the handover to the Soviets only two days away, the Mission had little time to explore, and there was no well-informed guide. They found hundreds of incomplete V 1s, and many spare parts for the V-2, even after the U.S. had spent the previous ten weeks since the factory was discovered stripping it before turning it over to the Soviets; as much as three hundred railway waggon loads of material may have been removed, in addition to numerous complete V-2s.

Fedden again: *'I knew several leaders of the German aircraft and engine industry before the war, and it was a sombre experience to meet them again after seeing the horrors of the slave camps like Nordhausen, with the incinerators for the corpses and the pitiful shambles of maimed and dying humanity. The Germans individually cannot escape responsibility for their collective actions which have consumed millions of innocent victims. For the most part these leaders of industry seemed quite unrepentant, and thought that it was bad luck that they lost after so nearly winning.'*

Between 20 and 24 June, the Mission examined BMW's plant at Munich and interviewed Dr. Bruckmann, Fedden's old friend, technical director of BMW's engine programmes; Dr. Amman, in charge of BMW's piston engine development; Mr. Willich, his top aide; and Dr. Sachse, senior engineer until 1942, and the man responsible for overseeing production of the 801 radial. While at Munich, the Mission examined several 801 developments, including a turbocharged version with hollow turbine blades; several of these were apparently abandoned at Kassel's airfield. They also examined the BMW 802, an '...*interesting and unorthodox design*', by Sachse; Fedden considered it '...*one of the most interesting piston engines seen in Germany*'. Of the BMW 803, designed by Dr. Spiegel of Siemens, the final report remarked, "Its layout and design appeared clumsy and rather indifferent." They also witnessed the BMW 003A1 run on the Herbitus test stand.

Unable to visit Daimler-Benz's Stuttgart works until the second trip in July, the Mission nevertheless found contradictions between his briefing before departure and what his interviews with the company's general manager, Mr. Haspel, and chief designer, Dr. Schmidt, told him. The Mission was informed that the 24-cylinder Daimler Benz DB 604 had been stopped in 1940 by RLM, and Daimler-Benz had considered 36-cylinder engines also.

At Völkenrode, the Mission found examples of Otto Lutz's swing-piston engine, developed co-operatively with Bussing of Braunschweig, with work also done by Mahle and Bosch. Fedden was dismissive. By contrast, he praised German use of wind tunnels in engine development, and suggested fuel injection was increasingly important for piston engines, especially as the number of cylinders rose.

The Mission's examination of fuel injection research was hampered by being unable to speak to the injection specialists at Junkers in Magdeburg or Deckel in Munich, but at Munich, a member of Bosch's development department, Dr. Heinrich, advised the Mission members that Germany had made few advances in the field beyond higher-capacity pumps, but learned that BMW had preferred the Bosch closed nozzle for the 801, while Junkers chose the open nozzle for the 213. They got better information on German aircraft spark plugs from BMW, Daimler, and Bosch, but not from Beru or Siemens, finding, in general, improvements focused on better performance at altitude or hotter, usually with better insulators or cooling.

Investigation of propellers found new research had been halted when RLM decided to focus on jets, only to be resumed. The Mission's interviews were limited to Vereinigte Deutsche Metallwerke (United German Metalworks, VDM), and, at Göttingen, they interviewed Dr. Stüper, who tested VDM's reversible propeller. This three-bladed unit had links to two electric motors, which could change the pitch at two degrees per second for constant speed or 60-100° per second for braking (in reverse mode); it was scheduled for production early in 1945, to be used by the Dornier Do.335, Dornier Do 317, and Fw.190. Mission members visited VDM's forging works at Heddernheim, on the outskirts of Frankfurt. Dr Eckert, of VDM spin-off company *Continental Metall Gesellschaft* (Continental Metal Company, CMG) confirmed Stüper's claims adding that CMG contemplated switching to hydraulic cylinders. CMG also had a project propeller that could eliminate engine overspeeding, by reducing pitch on two blades and increasing it on the other two. In addition, CMG had done some work on hollow propeller blades, one made from a simple rectangular tube with welded-on edges.

Fedden mused on what was the way forward: '*Far-reaching decisions face us over the future of the german scientists and research workers. Are we to use their vast accumulated knowledge for peaceful ends, or treat them as outcasts? The Russians as they often do seem to have made their own decision and acted upon it. I heard of one Bosch engineer, whose family I knew before the war, who had swam the Elbe to get out of the Russian zone; he said that the whole equipment of the Bosch*

Messerschmitt Me.262 111857 and a Junkers Ju.87 were discovered 'somewhere in the Munich area' in 1945. *[Hugh Jampton Collection]*

research headquarters at Spandau, with all their technicians, had already been removed complete to Russia.

The Americans have put tremendous drive into combing every aspect of German aeronautical technique and their greater manpower and unlimited transport facilities have enabled them to do the whole thing on a thorough and systematic plan.

Four members of the Mission went to Rosenheim and the BMW rocket development department at Bruckmühl, joined by Dr. Bruckmann, who informed them RLM had ordered rocket development which had begun in 1944. They were shown the BMW 109-718 assist rocket first. They were also shown the 109-558, used in the Henschel Hs 117 guided missile; the Report praised the 558, though mildly. In addition, the Mission was shown the 109-548, the sustainer motor for the Ruhrstahl X-4 (which Fedden described as an 'inter-aircraft rocket').

22 June, the Mission visited the Messerschmitt works at Oberammergau in the Bavarian Alps, where a Bell Aircraft team had already been working some five weeks; there was also a representative from De Havilland. Messerschmitt had taken over a former Heer barracks in 1943, and constructed 22 miles of tunnels. The works there were a taste of the future, and Fedden interviewed a number of Messerschmitt's senior engineers, including Hans Hornung, Joseph Helmschrott, and chief designer Waldemar Voigt. The Mission were shown examples of the Me-262, with Voigt blaming inexperience among pilots, and compressor stalls, for several accidents. They also saw a variety of projected designs, including the P.1101, P.1110, P.1112, and P.1108.

Most of the Mission returned to the UK on 1 July 1945, just Stern and Beeton remaining. On 4 July, they arranged for the Mission to see BMW's high-altitude engine test bench in Munich, which Fedden and three others did on 17 July, as part of an eight-day trip. A sub-group of the Mission returned to Germany from 16 to 25 July 1945, working out of Freising, to examine the BMW high-altitude test chamber at Munich, as well as German facilities at Stuttgart, Göttingen, Volkenrode, and Kochel. The BMW facility, codenamed Herbitus, was designed and operated by Christoph Soestmeyer, and was finished in May 1944. It was used for trials on BMW and Jumo turbine engines, as well as the 801; Soestmeyer reported that RLM had intended to build similar facilities at Rechlin, Berlin, Dessau, and Stuttgart. The test facility was a building 250 ft square and 70 ft high, containing a steel cylindrical altitude chamber some 12 ft in diameter and 30 ft long, with a detachable rear section, allowing engines to be wheeled in and out for testing. It was capable of altitudes of as much as 36,000 ft and test speeds of 60 mph. The Mission hoped to test the Derwent V and de Havilland Goblin at altitude. Fedden called it *'far in advance of any engine testing plant in England or America'*, and was sufficiently impressed as to suggest the facility be moved to Britain, but the Americans refused.

The Mission returned to the UK with a number of jet engines and rocket motors, turbine blades, and a large quantity of blueprints, totalling around 2,000 tons, yet this was much less than the Americans.

Unable to have the Herbitus test stand moved to Britain, and though Operation Surgeon had identified some desired 1,500 specialists to be brought to Britain, Fedden had no better luck persuading the UK Government to bring back German engineers and scientists, which both the Soviets under Operation Osoaviakhim and Americans were doing. In the event, only about 100 leaving Germany in 1946 and 1947 actually stayed in Britain.

Backfire

Operation Backfire came into being under the orders of the Supreme Commander, Allied Expeditionary Force, General Dwight D Eisenhower.

Following the successful conclusion of the tests, the War Office in London issued a huge 5-volume report detailing these operations. According to this report, the operation was *'...organised and conducted by the armies on the continent. Because it was to take place in the British Zone, the organisational set-up was predominently British. The actual work of building and launching the A-4 rockets was done by the Germans'.*

The race to discover and capture the secrets of the German missile began even before hostilities in Europe ended. Despite being an Allied military operation, it became something of a thinly disguised battle between the British and the Americans. The British had created Headquarters, Special Projectile Operations Group under the command of Major-General A M Cameron, British Army of the Rhine as a follow up to the myriad of technical reports already published by the RAE. On 20 May a request was submitted to G-2 Division (Technical Intelligence) Supreme Headquarters Allied Expeditionary Force for the allocation of parts for thirty complete rockets.

Four sources of supply were to be considered: the underground factory and assembly centre known as Mittelwerk at Niedersachwerfen near Nordhousen, the dumps of A-4 and associated ground equipment in both the 12th US Army Group and 21st US Army Group areas, and the various factories throughout Germany associated with A-4 production.

However, on 10 April 1945 the spearhead of the advancing American troops, Combat Command B (CCB) of the U.S. Third Armored Division, entered Nordhousen. Here CCB was to pause and link up with the U.S. 104th Infantry 'Timberwolf' Division before continuing its drive to the east. Several miles further, as they approached the foothills of the Harz Mountains, American troops discovered Dora and the entrances to the Mittelwerk tunnels. When walking into the first long tunnel, they were stunned to see railway freight cars loaded with V-2s. When word came of the incredible find, U.S. Colonel Holger Toftoy, Chief of Army Ordinance Technical Intelligence, immediately began arranging for 'Special Mission V-2' from his office in Paris. Its purpose would be the evacuation of 100 complete V-2s and specialised parts back to the United States. To support his mission, Toftoy had organised special rapid-response Ordinance Technical Intelligence teams attached to each Army group.

These teams were equipped with cameras, radios, transport trucks, and qualified personnel whose job it was to ferret out interesting weapons technology and record it. The team designated to investigate the Mittelwerk was headed by Major James Hamill of Ordinance Technical Intelligence. He was assisted by Major William Bromley in charge of technical operations and by Dr. Louis Woodruff, an MIT electrical engineering professor, as special advisor. The team was headquartered in Fulda, about 80 miles southwest of Nordhausen.

After rounding up captured German rolling stock and clearing a path into the tunnels, Special Mission V-2 succeeded in loading up and sending off its first 40-car trainload of V-2 parts and machine equipment on 22 May 1945. Nine days later, the last of the 341 rail cars left the Mittelwerk bound for Antwerp. Although the British properly protested that by prior agreement, half the captured V-2s were to be turned over to them, the Americans ignored these protests.

Sixteen cargo vessels bearing the components for a hundred V-2 rockets, finally sailed from Antwerp, destined for New Orleans and then White Sands. Hamill was not told that the factory would be in the Soviet zone of occupation. Consequently, quite a number of missile parts were left for the British and Soviets to discover.

Despite the indecent haste that the American armed forces demonstrated by riding roughshod over existing agreements in sharing the discovered technology, it was not until 15 March 1946 that they first static fired a V-2 at White Sands, and it was not until 16 April 1946, that the first V-2 was launched from the New Mexico missile range, some seven months after the British test-fired theirs! But that is getting ahead of the story somewhat.

To give an idea as to the scale of the Nordhousen facility, Operation Backfire documentation records that there were approximately 18 miles of underground tunnels and bays that had to be combed after the Americans vacated the site.

The British recovered 640 tons of equipment and components which they put onto rail cars for delivery to Cuxhaven where they intended to assemble and test fire a number of rockets.

Major Robert Staver from the Rocket Section of the Research and Development branch of the Ordinance Office was tasked in directing the effort to find and interrogate the German rocket specialists who had built the V-2. Since 30 April he had been in the Nordhausen area, searching the smaller laboratories for V-2 technicians. On 12 May Staver located his first V-2 engineer, Karl Otto Fleisher, who began to put him in touch with other Mittelwerk engineers who had not been part of von Braun's caravan to Bavaria. On 14 May Staver found Walther Riedel, head of the Peenemünde rocket motor and structural design section, who urged the Americans to import perhaps 40 of the top V-2 engineers to America. After their surrender to U.S. forces in Bavaria, Wernher von Braun's V-2 specialists were moved to a prisoner enclosure in Garmisch-Partenkirchen, where a variety of Allied interrogators questioned them. At this point the Americans had the missiles, they had the top scientists, but they were still missing the all-important Peenemünde documentation. Fourteen tons of Peenemünde documents had been hidden by Peenemünde engineer Dieter Huzel in an abandoned iron mine in the isolated village of Dornten in early April.

Von Braun had ordered the documents hidden to prevent their destruction by SS General Kammler, and to also use them as a bargaining chip in negotiating their fate with the Allies. As it happened, Karl Otto Fleisher was the only person remaining in the Nordhausen area who was aware of the general location of the V-2 documents hidden by von Braun's

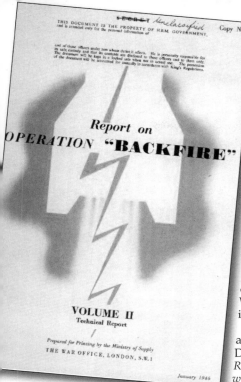

Operation Backfire documents

group. Staver tricked him into revealing the location of the papers on 20 May. In less than a week, the Dornten area was scheduled to fall into the hands of the British. A frantic scramble then ensued to transport the documents back to Nordhausen, where they were quickly shipped to Paris, and then to the Aberdeen Proving Ground in Maryland.

At the beginning of June, Staver requested that some of von Braun's senior engineers be sent to the area of Nordhausen to help identify which of the thousands of German technicians should be offered evacuation to the American zone before the scheduled handover of Nordhausen to the Soviets on 21 June 1945. On June 20, some 1,000 German V-2 personnel and their families were selected and gathered up, then placed aboard a long train, which eventually made its way to the small town of Witzenhausen, some 40 miles to the southwest and just inside the American zone.

Von Braun and his team were heavily interrogated and jealously protected from Russian agents. Dornberger later told British interrogators, *'The Russians sent one of my former engineers to me when I was with the Americans, who told me he had an offer to make on behalf of the Russians. We were to go back to Peenemünde and it would be rebuilt, along with a parallel factory in Russia, and they offered to pay us double what the Americans were offering us. We also could move our families with us, etc. We turned it down flat. Afterward, in the town of Witzenhausen, they tried to kidnap our leading men such as Dr. Wernher von Braun. They appeared at night as British soldiers in uniforms; I guess they didn't realise it was the American zone. Somehow they had obtained a proper pass, but the Americans quickly realised what was happening and sent them away. That's how those Russians operated, real kidnapping, they had no scruples at all'*.

Meanwhile, at the former Krupps gun-proving grounds at Cuxhaven all the material had been recieved by the British teams. Unfortunately, unloading and assembly was delayed as it was discovered that the demolition charges were still in place!

The British set up their first A-4 for launch on 1 October 1945, but during the pre-flight testing problems were encountered with getting the steam valves to open. The fault was traced to incorrect wiring of ground equipment and a second launch attempt was made at 1550 hours. The igniter functioned satisfactorily, and the flame pattern was good, but the main stage did not come on as the ground connector plugs did not throw off. German personnel on hand explained that this happened with about one in ten launches.

The rocket was re-set, a fresh igniter was inserted

One of the three V-2 missiles seen on its transportation trailer during Operation Backfire trials in October 1945.

and a second attempt to launch was made. This too failed, due to the ingniter being thrown out before it had time to light the fuel. A further attempt to fire was not made due to the failing daylight, and so the rocket was de-fueled - a process which took two and a half hours. The rocket was then sent back to the Vertical Testing Chamber for further overhaul.

The next rocket was set up for firing on 2 October and successfully launched at the first attempt at 1443 hours Zone A time. This was the first A-4 launching to be observed by Allied personnel at close quarters. Weather conditions were ideal; there was little wind and a clear sky. Only one fault was discovered during the main tests and this entailed the changing of an alternator. The behaviour of the rocket from the moment of take-off to the point of fall was perfect, the take-off was steady and the turn-over from the vertical occurred at the right moment. *Brennschluß* – or fuel cut-off occurred after 65 seconds.

The error in line was 1.2 kilometres left, and the error in range 1.9 kilometres short. The time of flight was four minutes 50 seconds. The method of control for range of this rocket and of all rockets used was by a time switch controlling cut-off of fuel. No form of radio control was employed.

The second launch took place on 4 October. Weather conditons were again ideal; there was no wind and visibility was almost perfect. The rocket, which had failed to fire on 1 October was set up, having been checked over in the intervening period, and was launched at 1415 hours. *Brennschluß* seemed to occur after approximately 35 seconds burning, and the projectile fell 24.95 kilometres from the point of launch, 0.99 kilometres to the left of the line of fire (2-28° deviation allowing for rotation). The time of flight was two minutes 16 seconds.

As the final report stated: *'It was originally thought that this failure was due to over-running of the turbine. If the turbine runs too fast there is a danger of explosion and air break-up. A centrifugal switch is therefore provided to give fuel cut-off if the rate of revolution of the turbine is excessive. In order that cut-off will not be given so soon as to land the rocket in friendly territory this switch is not energised until 40 seconds after launch. Cut-off can then be given in the hope that the rocket will reach enemy territory in one piece, even if it falls short. However, examination of the Kine Theodolite films (two stations followed the projectile through its complete course) indicated that thrust ceased after only 35 seconds. No satisfactory explanation has been produced. While the behaviour of the rocket was not satisfactory, its launching showed that it is possible to fuel and un-fuel a rocket, and later launch it, without carrying out any major overhaul'.*

The third and final rocket was launched on 15 October, as a demonstration to representatives from the United States, Russia, France, the Dominions, Whitehall, and the Press. Weather conditions were poor, there being low cloud and a 30 miles an hour surface wind ; but as the main object of the

Opposite page: details of the launch area, aiming point and radar coverage of the British 'Operation Backfire' V-2 launches on October 1945.

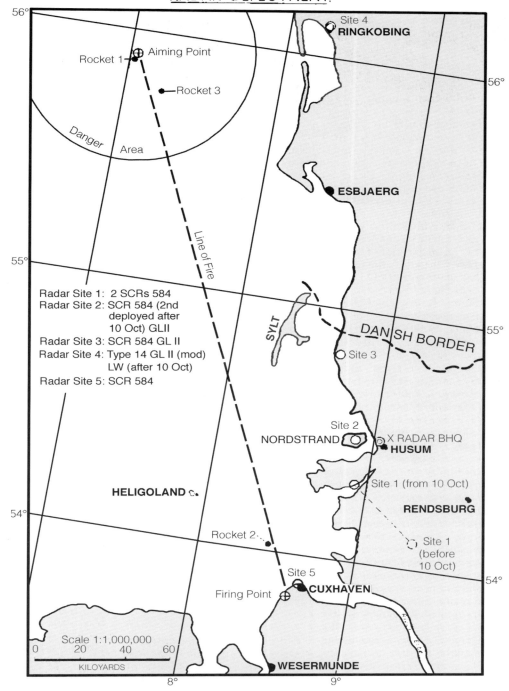

launch was to demonstrate to the visitors, it was carried through despite these poor conditions. The launch was made exactly on time, and no hitch occurred anywhere in the preliminary proceedings.

The rocket behaved normally, but fell 18.6 kilometres short and 5·3 kilometres right of the point of aim. The time of flight was four minutes 37 seconds.

For the record, it was not until 15 March 1946 that the Americans conducted their first V-2 static firing, and not until 16 April that they launched their first V-2 from the White Sands New Mexico missile range.

Dangers of the job at hand.
Recovering the German technology was one thing - safely investigating it was something completely different. Not only was there the ever present danger of hidden explosives set to either demolish the item if tampered with, or destroy it in the event of capture there was more subtle 'sabotaging' – deliberate or otherwise - of equipment than that to be considered. Documents from both the British and American archives are littered with reports of hidden explosive devices and signs of sabotage.

Much of the causes of this went back to the general mobilisation into the Wehrmacht from 1941 when the German aviation industry lost more skilled workers than it gained, and the proportion of unskilled workers, semi-skilled and 'forced labour' – a polite term for prisoners - increased.

Some firms, like Daimler-Benz Motoren GmbH, in contradiction to official policy, began employing women and unskilled labour to meet their personnel shortages in 1939. The longer the war continued the proportion of available skilled manpower continued to fall dramatically. At the Messerschmitt Kematen plant in the Austrian Tyrol the number of skilled workers fell from 57% in February 1941 to only 23% early in 1943. At Henschel the situation was even worse. The percentage of skilled labour dropped to only 11% by 1942. More dramatic was the great influx of foreign workers. By 1942 this was the main source of recruiting new workers in the aircraft industry. An increase of 100,000 workers between January and August 1942 consisted of 20,000 PoWs and 80,000 foreign labourers.

Ernst Friedrich Christoph 'Fritz' Sauckel (27 October 1894 – 16 October 1946) a German Nazi politician, Gauleiter of Thuringia and the General Plenipotentiary for Labour Deployment from 1942 until the end of the Second World War, sent 2.75 million foreigners to the Reich for work between April and November 1942 alone. 1,462 million of these were for the armaments industry. Thirty percent were women. Of the 2.75 million, 1.375

Two views of the erection and launch of one of the three V-2 missiles launched by the British during Operation Backfire in October 1945. This is throught to have been the 15 October launch for the Press and Allied officials.

The USSR sent six Soviet representatives, including Sergei Korolev, to witness the third, and last, British demonstration launch of a V-2 at Cuxhaven. The British had invited each Ally to send three observers to witness their test firing of a V-2. The Soviets sent six, not three. The British only admitted the allowed three into the facility to witness the test, forcing the other three to try to see, from outside, what was going on.

million were recruited in the occupied areas of the Soviet Union, 292,000 from Poland, 168,000 from France and 103,000 from Belgium. There were also 418,000 PoWs.

Initially the aircraft industry received mainly French, Belgian, Czech, and Dutch people, although later the industry had to be satisfied with the less skilled Russians and Ukrainians, etc.

By 1942 all the large aviation plants reported that foreign workers accounted for about 45% of the total workforce. However, the firms made every effort to keep as many skilled workers as they could in order to ensure regular production at acceptable quality. By 1 January 1943 the manpower in the airframe industry increased to 285,085 and in the engine industry to 167,245. For 1 July 1944 these figures were 342,601 and 202,999 respectively. The total aircraft industry employed 1.8 million men in December 1942, with a requirement for a further 1 million. The existence of such high requirements signified that manpower resources were becoming a serious problem. One reason was the fact that the RLM and the Luftwaffe always received fewer personnel than requested, especially in 1942 when low priority was given to aviation production. The OKW at this time preferred to direct manpower resources to the ground forces and armament production. Steps were taken to rationalise production and productivity rose sharply for the first time. The rationalisation of the industry was initiated by Milch. According to his reports airframe industry results increased by 40% during the period January to October 1942 and the engine industry by 65%. Productivity increased continually throughout the war. In 1944 the 'production miracle' happened. In this year the aviation industry, employing only 2.33 million workers, produced 40,593 aircraft, four times more than in 1940 with 1.2 million workers; and in 1944 as opposed to 1942 the manpower comprised mostly unskilled foreigners.

It is true that in 1944 production comprised mainly fighters of lower airframe weight, and quality had dropped. This was not because of the use of unskilled workers but also due to the many sabotage attempts by the foreign forced labour. The French workers were the main instigators of these actions. In the Heinkel Oranienburg plant the sabotage actions started at the drawing boards. Every worker made his own small 'action', wasting material, making useless or low-quality parts, or taking far longer over a particular operation than was allocated, assembling aircraft and engines with small hard-to-detect defects resulting in many serious and dangerous faults.

The Russians were more primitive in their sabotage attempts. In the Dora missile works for example, many of them would urinate into the delicate electronic boxes of the V-1 and V-2 flight control systems. Many were shot as punishment. However, as a result of these actions many lives were spared in England, which was the main target for both of these weapons, and also in Belgium, which was also targeted.

The firms mounted training programmes for the new workers to improve their skills, hoping to improve productivity and quality, but the slowly expanding workforce was rapidly declining in quality. A lack of

Dornier Do.335s in the final assembly hangar at Oberfaffenhofen, Germany not long after the site was overrun by advancing American troops in 1945. It was key sites such as this that Operation Lusty was structured to investigate. *[Hugh Jampton Collection]*

supervisory personnel also resulted in lowering of quality standards and an increase in sabotage activity. This explained why, for example, Messerschmitts built in Hungary using skilled Hungarian workers were superior in quality to the same models built in Germany using mainly forced labour via foreign workers. The RLM had been encouraging foreign licensed production in the 1941-1944 period, but after the summer of 1944, except for the Czechoslovakian plants, all this additional industrial capacity was lost. Hitler's demand in October 1943 that the aircraft industry be transferred to huge underground shelters in itself required much extra labour. Xaver Dorsh planned to build six very big underground plants, each 300,000 square metres, by November 1944. On 9 April 1944 Hitler demanded that Milch use 100,000 Hungarian Jews in the Todt Organisation to realise this plan, the final order being issued on 21 April. The Junkers plant alone was to produce 2,000 Junkers and 1,000 Me.262 aircraft per month.

The Jews were deported, but the organisation of the construction was very slow. Only one building was completed by the end of the war, using 10,000 Hungarian Jews working two twelve hour shifts. The work was exceptionally hard and life expectancy was very short. Finally the Reich gained nothing from the sacrifice of these workers.

Enter the Soviets
Up to 10 May 1945 the Allies had destroyed German industry to an estimated value of RM 320 billion. However, not all the aviation manufacturing facilities had been destroyed and the acquisiton started immediately. As the Reich toppled that spring, Josef Stalin was well aware of his opportunities to exploit German technical expertise in the interest of the post-war Soviet Military. 65% of the German aviation industry with its research and production facilities, including those of Austria and the Protektorat, fell into Soviet hands as the Red Army battled westwards. Technical plunder had first priority. As Soviet troops fought their way towards Berlin, a different kind of army trailed them. Although its members were riding army jeeps and carried pistols or submachine guns, their new oversised uniforms and lack of battlefield decorations betrayed recent civilians in them. These 'trade-union officers' as they were known in the Soviet Army, represented various Soviet industries charged with the task of locating and removing to the USSR machinery and equipment from the occupied Germany.

The activities of these Soviet 'trophy battalions' were officially started on 21 February 1945 by a decree No. 7563ss of the State Defense Committee, GKO. The document established permanent commissions at every Soviet Front (Army Group) occupying Poland and Germany and made these commissions responsible for the removal of industrial equipment and materials from both countries.

Within the 1st Belorussian Front, P. M. Zernov chaired such a commission, which also included A. N. Baranov and N. E. Nosovsky. The commission oversaw 80 engineers and scientists.

On February 25, 1945, Joseph Stalin signed a decree of the State Defense Committee, GKO, No. 7590ss creating a Special Trophy Committee within GKO. It included G. M. Malenkov, N. A. Bulganin, N. A. Voznesensky, A. V. Khrulev and Lt. General F. I. Vakhitov, the head of Chief Trophy Directorate. At the beginning of the effort, General Vakhitov led the force of 40 work battalions, which by September 1945 grew to 48 trophy brigades, 23 of which were deployed in Germany, seven in Poland and six in Czechoslovakia. According to the same decree, all members of trophy commissions at different fronts became representatives of the Special Trophy Committee.

One of the first Soviet groups, which would specialise specifically on rocket trophies, left Moscow on 23 April 1945. It was led by General Nikolai Petrov, director of the Scientific Institute of Airplane Equipment (NISO), and included Boris Evseevich Chertok from the NII-1 research institute. He was one of the several people who, in 1944, studied remnants of the A-4 rockets recovered in Poland. Although the official goal of the group was the search for avionics, radar equipment and aviation armaments, Chertok and others had been looking forward to learning as much as possible about the German rocket programme

All the aircraft manufacturing facilities in the Soviet-occupied areas – the Arado site at Babelsberg; ATG at Leipzig; Dornier at Wismar; Focke-Wulf at Marienburg; the Heinkel works at Rostock, Oranienburg and Schwechat; the Junkers site at Bernburg, Dessau; Siebel plant at Halle; the Henschel works at Erfurt and Berlin; Messerschmitt factory at Wiener Neustadt - all were stripped of their equipment. This included two of the world's largest hydraulic die

Following a devastating air attack by the RAF on Peenemünde on 17 August 1943, it was clear that missile assembly would need to be moved to protected underground sites. One such site was the former Wifo gypsum mine in the Kohnstein Hill on the southern border of the Harz mountains close to Nordhausen. The site was to be known as Mittelwerk (Middle Works). A private company was established on 24 September 1943 under the control of SS Brigadier General Hans Kammler, who had been responsible for building a number of extermination camps including Auschwitz-Birkenau. A sub-camp of Buchenwald was established within the Wifo mine in August 1943 in order to provide the workforce for the construction of the new underground facility, which later developed into the Dora (or Mittelbau) concentration camp this was eventually located in more traditional barrack accommodation close to the south entrance, but this wasn't completed until October 1944. Construction at the Mittelwerk site started in October 1943 with the driving of two parallel 'S' shaped tunnels right through Kohnstein Hill from north to south. The Mittelwerk facility consisted of the two main parallel tunnels.

A and B, each approximately 6,200 feet in length... The two parallel tunnels were 21 to 23 ft high and 29 to 36 ft wide. The cross tunnels were smaller in cross section with a total area for the complex of over one million square feet. A smaller service tunnel ran through much of the middle of the site between A & B tunnels. The southern section of Tunnel A and the first three cross galleries were used for V1 production. The middle section of the complex (gallery 21 - 42) was used for V2 assembly while the northern end (gallery 20 - 1) was used for Junkers aircraft engine production. Each of the main tunnels had two standard gauge railway lines running through it.

Above: V 1 production. Right: the main gallery of Tunnel A.
(all USAAF)

forging presses which had been used to produce the spars for the Junkers Ju.88, as well as machine tools and their models, equipment, materials, finished parts, drawings, and documentation – everything was removed.

Around June - July 1945, General Andrei Illarionovich Sokolov, led another group of specialists in Germany. Sokolov was the head of Directorate of Armaments of the Guards Mortar Units, (or simply Katyusha Units) as well as Deputy Artillery Commander on Reactive Weapons.

Sokolov's group included Yuri A. Pobedonostsev, M. S. Ryazansky, Eugene Boguslavsky, V. P. Barmin - all engineers, and Lt. Colonel Georgy Tyulin. The latter was a senior aide of the Chief of Scientific-Technical Department of the Chief Armaments Directorate of Guards Mortar Units, GMCh.

Tyulin took responsibility for accommodating and assigning positions within the A-4 search effort to the arriving Soviet aviation specialists within the headquarters of the Soviet military administration in Berlin.

On 3 August 1945, Gaidukov apparently received an official blessing from Stalin in the form of the Decree of GKO No. 9716ss, creating the so-called interagency commission on the A-4, which had power to recruit experts from various industries of the USSR.

As a result, on 8 August 1945, many of the former NII-1 colleagues, who in 1944 studied the Polish remnants of the A-4 rocket were called to the Central Committee of VKP(b) (Communist Party of Bolsheviks) on an individual basis, where they found themselves included in the Inter-ministerial Commission leaving for Berlin the next day. The engineers were advised to consider the assignment as a military draft. What 'the assignment' was all about should have been explained to them upon arrival in Germany. At the end of the briefing they were offered to ask questions. Accordingly, they had none.

Next morning, the engineers arrived at the Moscow Central Airport all dressed in oversized military uniforms embellished with shoulder-straps of majors and colonels.

In Berlin the engineers learned that restoration of a complete set of blueprints for the A-4 and restoration of an experimental production line for the missile were the primary goals of the commission, based in the German capital and led by General Kuznetsov. At the time the commission supervised Institute Rabe, but eventually embraced all facilities of former Mittlewerk, German industrial complex manufacturing A-4s, as well as the centre of development of anti-aircraft missiles and rocket control systems in Berlin.

Of many aviation specialists who arrived in Germany at the end of summer 1945, Mikhail Ryazansky, Viktor Kuznetsov, Yuri Pobenostsev, Eugene Boguslavsky and Zinovy Tsetsior joined Institute Rabe in Bleicherode.

Along with civilian engineers, the Chief Artillery Directorate, GAU, sent from Berlin to Bleicherode military officers and recent graduates of military academies, among them Georgy Tuylin, Yuri Mozhorin, Pavel Trubachev, Kerim Kerimov, all future prominent figures in the Soviet space program. During 1945, the total number of Soviet specialists in Germany working on the rocket programme reached 284.

Although the brightest Soviet specialists scoured postwar Germany in a search for rocket secrets, the overall effort still lacked expertise needed to unravel the intricacies of the A-4 design, particularly its complicated flight control system. Most key documents describing the flight control system were missing, available hardware was also lacking. In the meantime, Western allies of the USSR worked hard to assemble the creme de la creme of the German science and technology.

At the official level, the head of the Soviet aviation industry, A. I. Shakhurin brought the subject of German specialists in a letter sent to the Central Committee of the Communist Party on 27 June 1945. Shakhurin recommended the creation of a special regime bureau for the German aviation specialists, which would be run by the NKVD, the feared Soviet secret police. The organization would be essentially a design centre in prison, not unlike the Sharashkas, which had been the home of many Soviet engineers during 1940s.

The dismantling of the Arado Neuendorf plant began in autumn 1945. Order Nr 124 from Soviet Military Administration in Germany (SMAD – Sowjetische Militiiradministration in Deutschland) was the basis for the dismantling of the German military industry. The heavy plant roof constructions were transferred by sea to the Soviet Union; however, a number of vessels sank en route in the Baltic Sea due to severe overloading. The dismantling and transportation of the captured factories was supervised by special squads of engineers sent out from Soviet aircraft plants, and train loads of equipment were sent east. Much was destined for the lst Experimental Factory at Podberezhye, 100 km northwest of Moscow. Parts and plans from the Junkers Motorenwerke from Bernburg and BMW from Eisenach arrived at Krasnaya Glinka, just outside Kuybishev, for the 2nd Experimental Factory. Both factories were carefully established and equipped to serve as key centres for exploiting German engineers and specialists who were to work in close collaboration with experts of the Soviet Research Institutes and Ministry of Aircraft Production. The Second Experimental Factory, headed by Kuznetsov, using German and Austrian engineers, developed the world's most powerful turboprop engines for the famous Tupolev Tu-95 strategic bomber.

Back in Germany, Marshall Zhukov, the commander of the Soviet military administration, SVAG, issued an order, requiring the Chief of Workforce Department Colonel Ya. T. Remizov to form proposals on the employment of highly qualified German specialists by 31 August and to develop a compensation system for German citizens by the end of October 1945.

Other equipment and tools were distributed between different Soviet locations, or were used to create completely new factories. This happened with the Heinkel Rostock works, which were sent to Kiselovsk, 40 km from Stalinsk. This plant made large sub-assemblies for many Soviet heavy and medium bombers, starting with the Tupolev Tu-4 which was a direct copy of the Boeing B-29 bomber.

Soviet representatives in Germany launched a campaign to recruit German brains in aviation, nuclear research and rocket technology. Apparently, both stick and carrot methods were employed to get the right people. A Soviet-controlled radio station in the city of Leipzig reportedly promised good wages and personal safety to any Peenemünde veterans. In most cases, German engineers and qualified workers offered their services voluntarily, in exchange for decent salaries and good conditions by postwar standards.

In early August 1945, the representative of Special Committee within GOKO, M. S. Saburov informed Marshall Zhukov that around 1,000 German specialists worked for various Soviet research organizations and that number was expected to climb to 3,000. Saburov recommended a creation of specialized research organizations for respective Soviet industries. In response, Zhukov signed an order of SVAG No. 026 on 'Organization of work on using German technology by the Soviet industry'.

An absolute majority of German employees who joined the Soviet effort to restore the A-4 had no prior involvement in the programme, although the Soviets never considered work in Peenemünde a requirement. Among such individuals were Kurt Magnus, a first-class gyroscope expert and Dr. Hoch,

An Arado 234B sits in a hangar on a captured airfield at the end of the war. *[Hugh Jampton Collection]*

an avionics specialist. In October 1945, Dr. Blazig, a key specialist from one of the subcontractors in the A-4 program, joined Institute Rabe.

Operation Osoaviakhim was a comparable Soviet operation which took place with NKVD - The People's Commissariat for Internal Affairs or *Narodnyy Komissariat Vnutrennikh Del* - and Soviet army units 'recruiting' more than 2,000 military-related technical specialists from the Soviet occupation zone of post war Germany for employment in the Soviet Union.

Much related equipment was moved too, the aim being to virtually transplant research and production centres, such as the relocated V-2 rocket centre at Mittelwerk Nordhausen, from Germany to the Soviet Union, and collect as much materiel as possible from test centres like the Luftwaffe's central military aviation test centre at Rechlin, taken by the Red Army on 2 May 1945. The codename 'Osoaviakhim' was the acronym of a Soviet paramilitary organisation, later renamed DOSAAF.

The operation was commanded by NKVD deputy Colonel General Serov, outside the control of the local Soviet Military Administration which in a few cases, such as Carl Zeiss AG, tried to prevent the removal of specialists and equipment of vital economic significance for the occupation zone, unsuccessfully, as it turned out, with reportedly only 582 of 10,000 machines left in place at Zeiss. The operation took 92 trains to transport the specialists and their families - perhaps 10,000-15,000 people in all, along with their furniture and belongings. Whilst those removed were offered generous contracts (the specialists were told that they would be paid on the same terms as equivalent Soviet workers, which in post-war Germany was seen as a gain), there was little doubt that failing to sign them was not a realistic option.

Soviet troops captured some engineers who had formerly worked on the Arado Ar 234 jet bomber. They were taken to the Soviet Union, not returning to Germany until 1954. The documents that remained in 1946 on the Ar.234

The Arado 234 Blitz came in at least three different engine configurations, as a twin, with four individually podded engines or with engines podded in pairs. The Americans would later make use of podded engine data on a number of early civil and military aircraft designs. *(USAAF)*

The 'Category One' list contained a sub-list of German aircraft the Americans were desperate to obtain. Clearly the list had been worked up from a number of different sources - possibly an amalgamation of contemporary data from British and American - given the use of both 'undercarriage' (British) and 'landing gear' (American) in the description. Some machines were later discovered not to have existed or were radically different to the descriptions shown here.

had been given to the Ilyushin development construction department. Many design examples from the Arado Ar.234 jet can be found in a number of Ilyushin aircraft that followed, such as the Ilyushin Il-22, Il-28, Il-30, Il-46 and Il-54.

The major reason for the operation was the Soviet fear of being condemned for noncompliance with Allied Control Council agreements on the liquidation of German military installations. New agreements were expected on four-power inspections of remaining German war potential, which the Soviets supported, being concerned about developments in the western zones. The operation has parallels with Allied operations such as Operation Overcast, Operation Paperclip, and Operation Alsos, in which the Allies brought military specialists to the United States.

The start of Operation Lusty'

On 22 April 1945, the USAAF combined technical and post-hostilities intelligence objectives under an organization somewhat appropriately called the Exploitation Division with the code name 'Lusty', a codename derived, it is thought, from LUftwaffe Secret TechnologY. In recent years a myth appears to have grown up that 'Lusty' just applies to the recovery of a small number of Me.262s from Germany to the USA. This is far from the case - that was just one very small aspect of a much larger operation.

Five days later, on 27 April General Carl A Spaatz issued orders to form an Intelligence Section within the Air Technical Service Command with the specific task of scouring five countries for innovative Luftwaffe aircraft and equipment. Germany, Austria, Denmark Norway and France were to be searched.

Operation Lusty began with the aim of exploiting captured German scientific documents, research facilities, and aircraft. The Operation had two teams.

To preserve that scientific picture, the American teams boxed up

Item No.	Description	No. Reqd by Priority			Remarks
		I	II	III	
	Arado				
1	Ar.231	-	3	-	No details available
2	Ar.232	-	-	3	Transport/Cargo aircraft - 2 or 4 engines; tricycle undercarriage; might have additional 10 wheeled undercarriage.
3	Ar.234 Series A	3	4	6	Reported to have three skids for undercarriage; two jet engines (probably Jumo 004)
	Series B	3	4	6	Reported to have wooden wings and tail. Jumo 004 engines.
	Series C	3	4	6	Reported to have 4 BMW jet engines arranged in two pairs. (BMW 003)
4	Ar.240	-	3	3	Twin-engined, high altitude reconnaissance and fighter
5	Ar.393	-	3	3	No details available.
6	Ar.396	-	-	-	Dive bombing trainer.
7	Ar.420	?	3	3	Designation may have been confused with Ar.240. Also see Brandenburg-42.
8	Ar.432	?	3	?	Possibly a development of the Ar.232
9	Ar.440	?	3	?	Presumed development of Ar.240; not known how many exist. 2 DB.603 engines; bomber.
10	Augsburg-44	3	3?	5?	May be Me.109H; seen at Augsburg; had 45ft wingspan.
11	Brandenburg-42	?	3	?	An experimental aircraft of Arado design having a span of approximately 42 feet. May be the Ar.420.
	Blohm & Voss				
12	Bv.138	-	-	-	Three engined high wing flying boat.
13	Bv.141B	-	3	5	Single engined unsymetrical monoplane cabin and engine in seperate nacelles; fuselage extends to tail from engine nacelle only.

#	Name				Description
14	Bv.144	-	-	3	Twin-engined transport.
15	Bv.155	?	3	?	No details available.
16	Bv.222	-	-	3	Six-engined flying boat.
17	Bv.237	-	3	?	Single engined observation type; asymmetric tail plane.
18	Bv.238	3	?	?	Experimental six engined flying boat. Only one known to exist; span about 195 ft; may have Jumo.222 engines.
19	Bv.333	3	?	?	Reported (unconfirmed) as 4 engined bomber.
	Dornier				
20	Do.217	-	-	3	Twin engined bomber. Sub types M or P only.
21	Do.217P	-	3	3	High altitude version with 90 ft span with DB.605 engine in fuselage for supercharging wing engines (DB.603)
22	Do.235	-	3	?	Reported (unconfirmed) as 4 engined bomber; numbers also reported as applying to a Junkers aircraft.
23	Do.317	3	3	3	Experimental aircraft, not known how many in existence, reported as prototype of Do.217P, 2 DB.603 or 2 Jumo.213.
24	Do.335	-	3	3	Twin-engined multi-purpose aircraft. Two engines in tandem. One tractor propeller in front, one pusher propeller at rear. Experimental aircraft, not known how many exist.
25	Do.435	-	3	?	Reported as development of Do.335 with two Jumo.222 engines.
	Fieseler				
26	Fi.103	3	4	6	Believed to be a flying bomb; possibly 2 Argus-Rohr propulsion units.
27	Fi.156 (Storch)	-	-	-	Light reconaissance aircraft.
28	Fi.256	-	-	3	Light reconaissance aircraft, possibly development of the Fi.156 (Storch)
29	Flettner.282	-	3	4	Helicopter. Not known how many exist. Twin rotor; 550 hp Hirth engine.
30	Focke-Achgelis.223	-	3	4	Twin rotor helicopter powered by Bramo Fafnir engine. Not known how many exist.
	Focke-Wulf				
31	Fw.152 (Ta.152)	-	3	4	Scaled up Fw.190 with in-line engine. Reported for high altitude operation.
32	Fw.154 (Ta.154)	-	3	3	Reported to be the German version of the Mosquito.
33	Fw.183 (Ta.183)	2	(4)	(6)	Very fast, single engined jet engine.; experimental. Reported not complete.
34	Fw.189	-	-	-	Twin engined, tail boom. Obsolete.
35	Fw.190C	-	3	3	Number built unknown; may be increased wingspan.
	Fw.190 (BMW.801D) (Windei)	-	3	3	Jettisonable turbo supercharger.
	Fw. 190 (BMW.801E)	-	3	3	Fixed external supercharger.
	Fw.190 (DB.603)	-	3	3	Reported experimental; frontal radiator.
	Fw.190 (DB.628)	-	3	3	Reported experimental.
	Fw.190 V.13	-	3	3	Reported prototype. 40 ft span.
36	Fw.191	?	3	3(?)	Reported 2 engine; 70 ft span; pointed tail.
37	Fw.200	-	-	-	4 engine, low wing, military version of transport.
38	Fw.206	-	-	3	Reported 2 engine transport; believed not built.
39	Fw.237	-	?	3	Reported (unconfirmed) as multi-purpose; wood construction.
40	Fw.254 (Ta.254)	-	3	(3)	Presumed development of Fw.154. Experimental; not known how many exist.
41	Fw.300	?	3	?	Details not available.
	Fw.300A	-	?	3	Projected 6 engine bomber using BMW.801s. Also reported as Fw.400.
	Gotha				
42	Go.244	-	-	-	2 engine powered version of Go.242 cargo glider
	Heinkel				
43	He.111	-	-	-	2 engined obsolete medium bomber
44	He.111Z	-	-	3	2 He.111s joined, total of 5 engines.

#	Type				Description
45	He.115	-	3(?)	(?)	No details available.
46	He.162	3	4	6	High short wing; narrow track undercarriage; Heinkel-Hurth turbojet (109.011) mounted above fuselage; twin tail.
47	He.117A-3	-	-	-	Most common type; 4 engines coupled to drive two propellers.
	He.177A-5	-	-	-	Improvement on A-3
	He.177A-7	-	3	3	
	He.177B-5	-	3	3	103 ft span; 4 engines and propellers (probably DB-605)
	He.177B-7	-	3	3	Long span; otherwise similar to B-5
48	He.219	3	3	4	Twin engined night bomber tricycle undercarriage, 2 DB-603. Specially desired is the version with the jet propulsion booster unit slung under fuselage, and a version powered by Jumo.222 engines.
49	He.274	3	3	6	Turbo supercharged DB.603s armoured pressure cabin in nose. Prototyped by Farman, Paris - may also be found in Germany.
50	He.277	-	3(?)	3	Details unknown, probably development of He.177.
51	He.280	3	4	6	Twin jet propelled fighter.
52	He.343	3	3	4	Reported multi-seat fighter with 2 or 4 jet engines.
53	He.1060	?	3	4	No details available.
	Henschel				
54	Hs.126	-	-	-	Single engine, high wing, three place.
55	Hs.128	-	3	3	Experimental twin-engined high altitude aircraft. Not known how many exist.
56	Hs.129	-	-	-	Low wing, heavily armoured, ground attack type. 2 Gnome-Rhone engines.
57	Hs.130A	-	3	3	High altitude aircraft; probably 2 DB.628 engines.
	Hs.130C	-	3	3	High altitude aircraft; probably 2 BMW.801J engines with turbo supercharger.
	Hs.130E	-	2	4	High altitude aircraft with pressure cabin. 2 DB.603 engines 1 DB.605 in bomb bay to supercharge wing engines.
58	Hs.132	?	3	?	One place experimental fighter, possibly a jet.
	Junkers				
59	Ju.52	-	-	-	Three engined, obsolete transport.
60	Ju86R	-	-	3	Twin engined, high altitude aircraft with turbo supercharged Jumo-207 diesel engines.
61	Ju.87 (Stuka)	-	-	-	Obsolete dive bomber.
62	Ju.88H	-	-	3	Two engined bomber or fighter, fuselage lengthened over earlier sub-types.
63	Ju.88P	3	3	-	Fitted with 1-75mm or 1.88 mm gun.
64	Ju.90	-	-	-	Four engined, low wing transport aircraft.
65	Ju.92	-	3	3	Reported as bomber and troop transport; 4 Jumo-222 engines.
66	Ju.188	-	-	4	Twin engined development of the Ju.88.
67	Ju.248	3	4	6	Reported as Junkers version of the Me.163B; with tricycle landing gear; longer.
68	Ju.228	-	3	3	Experimental aircraft not known how many exist. Believed designed for Jumo-222 engines, but may be found with others.
69	Ju.388 J, K L	-	3	3	Believed to be fighter, bomber reconnaissance types respectively. Reported with turbo supercharged BMW.801, possibly pressure cabin.
70	Ju.390	-	3	3	6 BMW-801 or Jumo-213 engines, span 164 ft.
71	Ju.488	-	3	3	No details available.
72	Ju.588	-	3	3	No details available.
73	Ju.252	-	-	-	3 engined trtansport developed from Ju.52 - and other lines.; 'snoot' nose.
74	Ju.290	-	-	3	4 engined large transport, convertable to bomber. Developed

#	Name				Notes
75	Ju.352	-	-	3	from Ju.90. Possibly further developments of Ju.52.
76	Ju.452	-	-	3	Possibly further developments of Ju.52
77	Ju.488	-	3	3	This is actually the 4 engine Ju.188. 4 BMW-801 engines, 4 main wheels.
78	Ju.390	-	3	3(?)	Six engined transport. Production status not known.
79	Ju.588	-	3	3	Possibly confused with Ju.488.
80	Lechfeld-54	-	3	3(?)	Experimental development of Me.410 with a wingspan of approximately 54 ft.
81	Lechfeld-59	-	3	3	Experimental development of Me.410 with a wingspan of approximately 59 ft.
	Messerschmitt				
82	Me.109	-	3	3	Model with DB-628 engine (two stage supercharger) and model with MV50 (alcohol injection) desired. Also two place version (III)
83	Me.109/Ju.88 composite	-	-	-	Me.109 is mounted on top of Ju.88.
84	Me.110	-	-	-	Two engined fighter, now obsolete.
85	Me.209	-	3	3(?)	Developed to replace the Me.109, not in production; not known how many exist.
86	Me.3109	-	3	3(?)	Experimental aircraft; not known how many exist.
87	Me.161	-	3	3	Reported long range aircraft; four BMW-801 engines.
88	Me.162	-	-	-	Number now allocated to He.162 formerly bomber developed from Me.110.
89	Me.163	3	4	6	Tail-less single unit, rocket propelled aircraft.
90	Me.164	-	-	-	Reported as 2 engine transport development by Caudron; reported abandoned.
91	Me.208	-	-	3	Probably development of Me.108; communications type aircraft; tricycle undercarriage.
92	Me.209	-	3	4	Similar to Me.109; inward retracting undercarriage; developed as replacement for 109.
93	Me.210	-	-	-	2 engined development of Me.110; single fin.
94	Me.250	-	3	-	4 engined bomber; status unknown.
95	Me.261	-	3	3	Reported as single seat fighter and dive bomber with two DB-603 engines.
96	Me.262	-	4	6	Twin unit jet propelled fighter. One aircraft already shipped to Wright Field.
97	Me.263	2	4	6	Reported as development of Me.163.
98	Me.264	-	3	3	Four engined experiemental; not known how many in existence; one known to have been fitted with Jumo 211s.
99	Me.309	-	3	4	Intended as replacement for Me.109; more advanced than Me.209; tricycle gear.
100	Me.310	-	-	3	Presumed development of Me.210.
101	Me.323	-	-	-	6 engined powered version of Me.321 glider.
102	Me.324	3	4	6	Reported about Me.109 size; two Argus-Rohr jet engines; two spring skids.
103	Me.328	3	4	6	Reported as single seater; rocket propelled; 20 ft span.
104	Me.409	-	3(?)	3(?)	Presumed further development of Me.109.
105	Me.410	-	-	-	Development of Me.110.
106	Rechlin-60	3	?	?	In general similar to Me.110, but greater wing span (about 60 ft). More taper on trailing edge than on leading edge.
107	Rechlin-66	3	3	?	Experimental aircraft of span approximately 66 ft seen at Rechlin. Sharply swept forward wing and long thick nose. Possibly a tail first aircraft. May be Ju.287.
	Siebel				
108	Si.204	-	-	-	Small, 2 engined trainer.
109	Si.304	-	-	-	Probably development of Si.204.
110	'Volksjaeger'	X	X	X	Refers to single engine jet fighters in general.
111	Z.SO.503	-	-	3	Designed by S.N.C.A.S.O. and Zeppelin based on Me.323 but larger; span 230ft; twin fins and rudders; has been reported as Me.523

everything they could and immediately shipped it to Wright Field, Ohio. Inter-service rivalry still occurred of course. First on the scene at one location, Navy exploitation teams quickly boxed up the hardware and technical data in large crates and labelled them 'US Navy'. Two days later, Army teams made it to the same location, whereupon they crated the Navy boxes in larger crates and relabelled them 'US Army.'

Team One, under the leadership of Colonel Harold E. Watson, a former Wright Field test pilot, collected enemy aircraft and weapons for further examination in the United States. Team Two 'recruited' scientists who collected documents and investigated facilities.

A document – almost a complete manual - of what was wanted, and how the teams were to go out and locate and handle the recovery was prepared by the Director of Technical Services, Headquarters Air Technical Services Command in Europe in May 1945. This was published as the *'Category One List – Enemy Equipment Desired By Wright Field'* and replaced an earlier 'Category 'A' List issued that previous February.

Under the heading 'SECRET' it went on to explain the objectives, what was required and how these items were to be selected, and deserves looking at in detail:

1. *The chief objective of Air Technical Intelligence activities in Germany, at the present stage of the war, is to obtain any and all information which may be applied to the prosecution of the war, against Japan. There is good foundation to the believe that the Germans have made many, if not all, of their developments available to the Japanese. Hence, knowledge of German development is vitally needed to:*
   a, Devise effective countermeasures, both technical and tactical.
   b. Insure the technical superiority of our own equipment.
2. *This list of enemy equipment desired for technical intelligence and research purposes, is published as a guide for those engaged in gathering German materiel for intelligence purposes, It is a revision of, and supersedes, the Category A List of Enemy Equipment Desired by Wright Field, published in February 1945.*
3. *Two types of enemy materiel are listed - that which is known to exist or has been reported, and that on which no specific information is available. This latter type is listed in general terms. Specimens of German versions of these types of*

Two men from the 7th Army 'look under the hood' to inspect the compressor and Reidel starter-motor of one of the Junkers Jumo 004s of this Me.262 found abandoned at Giebelstadt, near Frankfurt. *(USAAF)*

equipment are required for examination and evaluation.
4. As it is of the utmost importance to Wright Field to receive, promptly, specimens of new German equipment, particular attention is invited to the method of scheduling priorities for shipment as follows:
a, Priority I Q will be shipped by air with the least possible delay. Consolidated shipments will be accompanied by a courier.
   1. In general, this priority applies to experimental models, the first captured specimens of production items, or significantly modified standard equipments which are considered to be capable of contributing valuable information as outlined in para 1a above.
   2(a). The following general classes of German equipment are considered to rate Priority I:
      (a) Jet or rocket propelled aircraft and accessories or instruments peculiar to them.
      (b) Jet or rocket propulsion engines and accessories or instruments peculiar to them.
      (c) Experimental high powered reciprocating engines and accessories.
      (d) Gas turbine engines combined with propeller drive.
      (e) All controlled missiles, especially the control mechanisms.
      (f) Methods of improving aircraft, engines, or crew performance at high altitudes (including cabin supercharging, oxygen equipment etc).
      (g) Methods of improving night operation of aircraft and crew (including exhaust flame dampers, night vision aids etc).
      (h) Special fuels, lubricants, anti-icing, finishes, alloys.
      (i) samples of advanced or new manufacturing techniques.
      (j) new or improved guns, rockets or rocket launchers.
      (k) advanced or experimental photographic equipment, notable high altitude aerial cameras, gun sight cameras, special lenses.
      (l) Microwave radar, anti-jamming arrangements, infra-red and associated equipment, radio control gear, television, any novel electronic equipment.
      (m) Any standard item showing significant improvements applicable to the distinct advantage of our own equipment.
   (b) Priority II - equipment to be expedited but does not require courier or air transport, except where surface transport facilities are inadequate.
(1) In general, this priority applies to the following two classes of enemy equipment.
   (a) A limited number of equipments of the same type as, but in addition to, that shipped under Priority I. This shipment is required to assure a sufficient number of the items shipped under Priority I to continue and expand the investigations prompted by the first received items. This will explain why many items list numbers required under more than one priority.
   (b) Equipment whose relative importance as outlined in para 1 does not warrant Priority I, but which is required for prompt investigation.
(c) Priority III – routine shipment
   (1) Material not required for immediate investigations , quantities of that Shipped under Priorities I or II Priority III. This will provide materiel for later investigations.
(d) It will be noted that 3 designations are given in the list for quantities. Those are explained as follows:
   (1) Numbers are given when information is sufficient to permit specific quantities to be requested.
   (2) 'X' indicates that the subject warrants the importance of the priority but no specific items are known. Samples of each type found are required. Quantities required may be judged by comparison with other items, bearing in mind the size and importance of the item in question.
   (3) A question mark (?) indicates that available information is insufficient to

make the priority certain. Examination of any recovered specimens in the light of the above explanation of priorities should be made to determine whether the indicated priority should be changed.

Certain items, particularly aircraft and engnes, are listed but not required. These have either been previously examined or are otherwise of no further interest .

The manual went on to describe a set of general instructions for the guidance of personnel 'in the field':

1. It is important that materiel shipped under Priority I or Priority II have any available information as to where found, how used, what used with, etc., securely attached to it. This is often a very great aid in proper evaluation.
2. Because of the incompleteness of our knowledge of German devolopments, field personnel will be required to evaluate much materiel found in the light of the fore-going definition of priorities. It is important that the existence of Priority I materiel be made known immediately in order that arrangements for its prompt inspection and removal can be made.
3. The existence of factories or experimental establishments dealing with Priority I Items should be reported immediately in order that arrangements for guarding, inspection and evaluation by qualified technical specialists may be made. Factories containing exceptionally large or otherwise noteworthy machinery should also be reported at once.
4. It is important that special tools , manifolding, wiring such as ignition harness, electric and hydraulic connectors and couplings etc., be searched for and sent with all items recovered to make complete examination and test possible.
5. NOTE: In dismantling aircraft, extreme care should be taken in separating and properly tagging all electric and hydraulic lines. If possible control surface cable tensions should be measured before dis-assembly and cables then properly tagged.
6. All stocks of German-captured Russian material should be reported at once.

The manual then laid out specific reporting instructions and eleven classes of items the Director of Intelligence was interested in:

All personnel having access to this list of equipment are requested upon locating any listed item to signal all pertinent details to the Director of Intelligence, USSTAF, Attention: Exploitation Division.

Arrangements should be made to have the item guarded. The D/I will then take the necessary action to have the article inspected or removed. If the item has previously been obtained in required quantity D/I will issue instructions to the

Me.262s were discovered in various conditions, parked on hardstands or in revetments. This 262, parked on the edge of a forest clearing is missing a number of items, including engines. (USAAF)

reporting agency to release the article for disposition in accordance with SHAEF or Comm Z directives.

Class	1	AIRCRAFT
	1A	Complete aircraft and/or airframes
	1B	Airframe accessories

(This section includes all equipment built into or onto the airframe but not directly a part of the propulsion system or other classes dealt with separately).

Class	2	PROPULSION
	2A	Jet engines
	2B	Reciprocating engines
	2C	Propellers
	2D	Take-off assist devices
	2E	Cooling Systems.
	2F	Exhaust systems
	2G	Engine mountings
	2H	Fuel systems
	2I	Ignition systems
	2J	Oil systems

Class	3	EQUIPMENT
	3A	Aero-medical equipment
	3B	Air Transport equipment
	3C	Emergency rescue equipment
	3D	Ground handling equipment
	3E	Personal equipment.

Class 4		COMMUNICATIONS AND RADIO NAVIGATIONAL AIDS
	4A	Airborne communication sets.
	4B	Airborne radio navigational aids.
	4C	Countermeasures.
	4D	Ground radio equipment.
	4E	Interphone systems
	4F	Miscellaneous equipment

Class	5	RADAR
	5A	Airborne radar.
	5B	Ground radar
	5C	Gun-laying radar.
	5D	Radar countermeasures.

Class	6	INFRA-RED
	6A	Picture resolving devices (viewing devices).
	6B	Miscellaneous aircraft identification devices

Class	7	SPECIAL WEAPONS
	7A	Aerial torpodoes.
	7B	Flying bombs.
	7C	Guided missiles

Class	8	ARMAMENT, ORDINANCE AND CHEMICAL WARFARE
	8A	Ammunition
	8B	Ammunition accessories
	8C	Bombs

One of the Me.262s captured at Giebelstadt, near Frankfurt after General Patch's 7th Army had overrun the airfield. *(USAAF)*

8D	Bombing accessories
8E	Bombsights
8F	Chemical Warfare
8G	Guns and cannons
8H	Gun accessories
8I	Gunsights
8J	Turrets
8K	Pyrotechnics

Class	9	PHOTOGRAPHIC
Class	10	INSTRUMENTS
	10A	Automatic pilots
	10B	Engine Instruments
	10C	Flight Instruments
	10D	Navigational Instruments
	10E	Miscellaneous Instruments

Class	11	MATERIALS
	11A	Adhesives
	11B	Finishes
	11C	Fuel and Lubricants
	11D	Metals
	11E	Plastics
	11F	Woods
	11G	Miscellaneous

Under the leadership of Colonel Harold E. 'Hal' Watson a team of pilots, engineers and maintenance men, were assembled used the lists generated at Wright Field to collect aircraft.

Harold Ernest Watson was born in Farmington, Conn., on 19 November 1911. He graduated from high school there in 1929, and four years later received a degree in electrical engineering from the Rensselaer Polytechnic Institute at Troy, N.Y. That September he joined Pratt & Whitney Aircraft Company as a research engineer.

Appointed a flying cadet on 15 February 1936, Watson graduated from Advanced Flying School at Kelly Field, Texas, a year later. Assigned to the 96th Bomb Squadron at Langley Field, Va., the following year he attended the Air Corps Navigation School there. Moving to Ohio in November 1939, he performed research development and procurement work, and later was named chief of the Quality Control Division. In 1941 he received his master's degree in aeronautical engineering from the University of Michigan. Watson became one of the few USAAF pilots to have experienced jet-powered flight when he operated the turbojet-powered Bell P-59 Aircomet from Wright Field.

Going overseas on 17 September 1944, Watson was director of maintenance in the Ninth Air Force Service Command attached to the Supreme Headquarters Allied Expeditionary Force (SHAEF) in the European Theater of Operations. Colonel Watson was no stranger to flying new or unusual aircraft, and his engineering background made him a natural choice to lead the Air Technical Intelligence effort. Since the target lists he had been given included aircraft of all types, Watson divided the effort between two teams. Group One under the command of Captain Fred B McIntosh was dispatched in search of piston-engine aircraft and non-flyable jet and rocket equipment, while the other team commanded by Lt. Robert C Strobell was charged with collecting flyable jet aircraft, mainly from the former Me.262 base at Lager Lechfeld, twelve miles south of Augsburg.

Watson had been given what was termed an 'Eisenhower Pass' which

The 'Category One' list not only recorded airframes the Americans were interested in - it continued with a list of engines, propellers....

Item No.	Description	No. Reqd by Priority I	II	III	Remarks
	Jet Propulsion engines				
1	Argus-Rhor	2	20	20	Reported constructed from sheet iron; only usable ones.
2	Argus jet engine	2	20	20	
3	Athodyd units	2	20	20	Experimental type of unit. Probably not in production. Commonly known as 'ram-jet'
4	BMW 003	2	20	20	Reported to burn gasoline, may be used in He-280. Similar to Jumo-205.
5	He-58A	2	20	20	Reported used in He-280; Campini type (?)
6	Hirth jet engine	2	20	20	Details not available.
7	Impulse duct	2	20	20	Only development beyond V 1 flying bomb type.
8	Junkers TL (Jumo 004)	1	20	20	Used on Me.262.
9	R-211	2	20	20	Used in Me-163. Reported made by Walter of Kiel.
10	Any new development engines or mods of existing models propeller engine.	2	20	20	All jet engines are of the utmost interest, of jet including turbine propeller engines. German designations: TL - Turbo jet engine. PTL - Turbine driven
	Reciprocating engines				
1	Bramo (Fafnir) - 323	-	-	-	
2	BMW-132	-	-	-	9 cyl air cooled radial
3	BMW-801	-	2	-	14 cyl radial. Later models desired (D, G, and later), where significant changes are evident (2 stage or turbo supercharger etc)
4	BMW-802	-	2	2	Reported 2 row, 18 cyl radial and in production.
5	BMW-802	2	2	4	Reported 2 row, 14 cyl radial; driving contra-rotating props; 4,000 hp.
6	BMW-804	-	2	2	Experimental engine, not known how many exist. Reported 27 cyl radial (three banks of 9 each) 2,800 hp.
7	BMW-805	2	2	4	Reported 3 bank radial; 3200hp; status unknown.
8	BMW-806	2	2	4	Reported 28 cyl radial; similar to BMW 803.
9	BMW-816	1	1	-	Reported experimental works; abandoned; description unknown.
10	DB-601	-	1	2	12 cyl inverted V, liquid cooled; desired if with 2 stage or turbo-supercharger.
11	DB-603-603 AS	-	1	2	12 cyl inverted V, liquid cooled; desired if with 2 stage or turbo-supercharger. DB-603 AS
12	DB-605	-	1	2	12 cyl inverted V, liquid cooled; desired if with 2 stage or turbo-supercharger
13	DB-607	-	1	2	Experimental engine, not known how many exist. Reported 24 cyl compression ignition.
14	DB-608	-	1	2	Details not available.

15	DB-609	-	1	2	Reported as 16 cyl, inverted V, liquid cooled.
16	DB-610	-	-	2	Two DB-605s couple to drive one propeller.
17	DB-612	-	1	2	Reported as DB-601 with redesigned cyl heads and rotary valves.
18	DB-613	-	1	2	Two DB-603s couple to one propeller, possibly contra-rotating.
19	DB-614	-	1	2	Experimental engine, not known how many exist.
20	DB-620	-	1	2	Details not available.
21	DB-623	-	1	2	Reported as development of DB-603 with addition of turbo-supercharger.
22	DB-625	-	1	2	Reported as development of DB-605 with addition of turbo-supercharger.
23	DB-626	-	1	2	Reported as development of DB-603 with addition of turbo-supercharger.
24	DB-627	-	1	2	No details available.
25	DB-628	-	1	2	DB-605 with two stage supercharger and intercooler.
26	DB-632	-	1	2	No details available.
27	Jumo-205C	-	-	-	6 cyl, opposed piston, diesel, liquid cooled.
28	Jumo-207, 207E, J	-	-	-	May be equipped with turbo-supercharger, similar to Jumo 205.
29	Jumo-210	-	1(?)	1(?)	details not available.
30	Jumo 211	-	1	2	12 cyl, inverted V, liquid cooled, mechanically driver supercharger. Desired with 2-stage supercharger only.
31	Jumo-212	-	1	2	Experimental engine, not known how many in existance. Thought to be 24 cyl 'X' or two Jumo-211 units side by side.
32	Jumo-213	(1)	1	2	12 cyl, inverted V, liquid cooled. Reported version with 3-stage supercharger particularly desired (Priority 1)
33	Jumo-222	-	1	2	Experimental engine, not known how many in existence. Thought to be 24 cyl engine made up of 6 blocks of 4 cyl arranged radially. Possible combination of liquid and air cooling.
34	Jumo-223	-	1	2	Experimental engine. Not known how many in existence.
	Propellers				
1	Feathering pumps & motors	-	-	X	Samples
2	Escher-Wyss Prop	X	X	X	Swiss reversible pitch propeller, experiemented with by Germans.
3	MeP-8	1	2	2	Reversible pitch propeller.
4	New type propellers	1	2	2	Particularly for contra-rotating types.
5	Propeller controls	1	X	X	Particularly for contra-rotating types.
6	Propeller governors	1	X	X	Particularly for contra-rotating types.
7	Prepeller hubs	1	2	2	Particularly contra-rotatinhg types and reversable pitch types.
8	Hollow metal blades	X	X	X	Experiments with 2 types of hollow steel blades reported.
9	Unusual blade construction	X	X	X	Plastics, wood greater than 16 feet dia, hollow aluminium etc.
10	Stamping process for propeller blades	-	X	X	Location of machinery and details important. Samples of blades desired.
11	VDM Propellers	-	X	X	Experimental models, samples.

basically allowed him to travel anywhere and acquire anything on behalf of the USAAF. The document - in English, French and German - stated he was allowed:

  1)   To examine or remove any captured aircraft or item of enemy air or Radar equipment whether found in the field, or in factories, workshops, stores or dumps.
  2)   To carry a camera for the purposes of photographing such equipment as he deems necessary.
  3)   To travel anywhere in the zones occupied by the Allied forces.

All service authorities are to assist the officer in the discharge of his official duties in every way, including the provision of petrol, oil rations and accomodation, should they be required.

Watson met a veteran P-47 pilot assigned to the staff there by the name of Lieutenant Robert C. Strobell while assigned to the 1st Tactical Air Force headquarters. Although their duties rarely brought them into contact with one another, the two did have a rather odd opportunity to share a cockpit on one occasion. In early 1945, Watson received a request to fly back a stricken B-17 that was several miles away in France. Knowing that Strobell was a seasoned aviator, the Colonel made it a point to gather up Strobell on his way out of the door. Of course, the young fighter pilot had no great ambition to lumber about in a damaged bomber, but after a harrowing flight the two successfully recovered the machine.

The Lieutenant's performance on that day clearly impressed Watson, for a few months later when word came down to assemble the exploitation teams, Strobell was immediately summoned to direct the efforts of the jet recovery group.

On 20 May, 1945, Strobell received orders assigning him to the mission. He recalls the meeting: *'Watson came into my office with a stack of documents on the Me.262, and simply told me to draw field gear, go to Lechfeld, teach mechanics to restore the Me.262 to flying condition, teach pilots to fly the jet, and prepare to ferry the jets out of Germany. The whole meeting lasted less than two minutes. I*

ME262A-1A 'V083' wearing the legend *'Feudin 54th AD Sq'*. This machine - the deveopment aircraft for the 50mm canon installation was later named *'Wilma Jeanne'* and then *'Happy Hunter II'*. (USAAF)

told him that I was delighted. He didn't bother to ask if I had any questions. Neither of us knew how to operate or fly the Me.262, and so there were no answers.

The intelligence reports that I had been given indicated that there were Me.262s on the field that could be restored to flight condition. At that time I understood that there was a German crew at Lechfeld working on the jets, but I had no knowledge of how many jets were on the field'.

Strobell was wary of what he would find there, as the area had fallen to the U.S. Army only two or three weeks before. While the terms of a surrender were in place, many pockets of resistance were rumoured to be active, especially in southern Germany and Austria – known to be Me.262 country. Lechfeld was still very much regarded as 'enemy territory'.

American Army units had moved into the Lager Lechfeld area in early May 1945 during the Western Allied invasion of Germany and seized the airfield with little or no opposition. Initial reconstruction plans for the base to be used as a United States Army Air Force field were cancelled after the German capitulation on 7 May, and the facility was garrisoned by United States Army units, although United States Army Air Force personnel were sent to the base to evaluate the Messerschmitt aircraft left at the airfield. It was designated as Advanced Landing Ground 'R-71'.

Also under Colonel Watson's command were the 54th and 56th Air Disarmament Squadrons, and the 2nd Air Disarmament Wing (Provisional). Before the arrival of Watson's Air Technical Intelligence (ATI) team, Platoon No 1 of the 54th ADS was ordered to Lager Lechfeld to prepare any suitable Me.262s discovered for eventual flight testing by the American pilots. Watson had also organised the selection of a number of former Messerschmitt mechanics to provide skilled support to the American units.

Watson's Whizzers

It was not long before the jet team gained the nickname of 'Watson's Whizzers' as they were essentially the first jet unit in American military aviation history.

When No. 1 Platoon of the 54th ADS, commanded by Master Sgt Eugene E. Freiburger, (some accounts spell his name as 'Frieburger) arrived in the Augsburg area on 29 April, they first moved into the Messerschmitt office building on 1 May and set up their squadron there. A team was then dispatched to the airfield at Lechfeld, where they found the infantry, on moving through the area, had smashed and damaged innumerable aircraft, leaving many totally destroyed.

Brigadier General Harold Ernest Watson *b.* 19 November 1911. *d.* 5 January 1994.

Ordered by Watson to prepare fifteen Me.262s for flight, Freiburger discovered that several of the aircraft had been booby-trapped with small amounts of TNT underneath each pilot's seat. The platoon's booby-trap expert, Staff Sgt Higgins from Birmingham, Alabama, made an assessment of each aircraft to ensure that they were safe to be handled. This accomplished, a bomb service truck towed each aircraft up towards a partially damaged hangar which had been converted into a workshop.

The damaged parts from each aircraft were removed and the German mechanics, together with the two American mechanics, reassembled them into complete, airworthy machines.

The Americans discovered that the fuselages could be split into three sections, each being interchangable with others. All they had to do was replace damaged sections with good items. Some damaged components were replaced by stripping Me.262s stored in revetments off the autobahn which had been used as an airstrip for the airfield of Fürstenfeldbruck - located between Augsburg and Munich - and which only days before had been used as a fighter strip by the Luftwaffe.

Somewhat surprisingly, the working relationship between the two former enemies was remarkably good, as Eugene Freiburger recalled: *'All in all the quality of work that the German mechanics did was good. The relationship that existed between all of the people and us was on a good basis, for we all inderstood what we were trying to do and I had no complaints. We kept a check to see that things were going right but as two of their own pilots were going to be flying the aircraft at first, the German mechanics were keen to do the best job possible.'*

With the arrival of Peace in Europe on VE-Day, the Whizzers soon added Luftwaffe test pilots to the team. One was Hauptman Heinz Braur. On 8 May 1945, Braur flew 70 women, children and wounded troops to Munich-Riem airport. After he landed, Braur was approached by one of Watson's men who gave him the choice of either going to a prison camp or flying with the Whizzers. Braur thought flying preferable.

The 'Category One' List contained a whole range of specialist items of equipment.

Item No.	Description	No. Reqd by Priority			Remarks
		I	II	III	
	Take-off Assist Devices				
1	Liquid fuel jet units				Application to jet aircraft or new type desired.
2	Solid fuel rocket units	12	20	20	Particulalrly used with jet planes.
	Cooling systems				
1	Coolant pressure relief valves	-	-	X	Samples
2	Coolant Systems	-	-	X	New types and developments including coolers, de-aerators and regulation of temperatures.
3	Cowling attachments and arrangements	-	-	X	Including automatic and normal regulators of cowl flaps and other temperature controls
4	Hose and Hose Clamps	-	-	X	Samples of new types.
5	Intercooler (supercharger)	-	X	-	Any type; samples.
6	Oil cooler surge protection valves	-	X	X	Samples of all types for existing engines desired Also samples of experimental types
	Exhaust Systems				
1	Exhaust collectors	-	X	-	Types apparently using special materials and arrangements desired particularly.
2	Exhaust gas heat exchangers	-	X	-	Particularly details of use for wing de-icing.
3	Flame damping exhausts	-	X	-	Any new type.
	Engine Mountings				
1	New shock mountings	-	X	-	Vibration isolating amnd quick disconnect mounts and fittings all of particular interest; samples desired.
2	'Power Egg' types	-	X	-	Details of arrangement; samples if applied to large engines.
	Fuel Systems				
1	Automatic engine control units	X	X	-	Particularly complete details on single lever controls for all operation of all engine controls.
2	Direct Fuel Injection	X	X	X	Samples of all types for existing engines desired. Also samples of experimental types.
3	Electric and hydraulic control units	-	X	X	Samples required.
4	Fuel Filters	-	-	X	
5	Fuel Pumps	-	-	X	
6	Fuel Condensers	-	-	X	

#	Item				Description
7	Fuel System Coolers	-	-	X	
8	Fuel System Valves	-	-	X	
9	Fuel Vent Systems	-	-	X	
10	Power Boost at Altitude	1	2	4	GM-1 Device for adding oxygen to engine at altitude, source being liquid oxygen, or liquid nitrous oxide carried in tanks.
		1	2	4	MV-50 Corresponds to our water injection device.
11	Turbo Installations	X	X	X	Particularly turbo regulators.
12	Turbo superchargers	X	X	X	Samples of all types desired.
13	Turbine Wheels	X	X	X	All types - particularly those for jet engines for test, and replacement on jet engines required.
14	Water Recovery Systems	X	X	X	For recovering water from exhausts for water injection etc.

Ignition

#	Item				Description
1	Ignition Systems	X	X	X	Particularly low-tension systems ie low tension distribution with individual high tension coil for each cylinder.
2	Ignition Cable	-	X	X	Samples from high altitude engines particularly.
3	Ignition syst pressurization	X	X	X	Pumps, harnesses, accessories etc.
4	Low Tension starting vibes	-	-	X	Samples
5	Magnetos	-	-	X	Samples.
6	Spark Plugs	-	-	X	Samples.
7	Spark Plug Cleaning	X	X	-	Reported systems are installed (attached) on engine for the purpose of cleaning spark plugs while in flight.

Oil Systems

#	Item				Description
1	Booster pumps in external oil system.	-	X	-	
2	Oil Systems	-	X	-	Any experimental types and developments including new methods of breathing and means of reducing foaming at altitude.
3	Oil system de-aerators	-	X	-	
4	Oil system pressure relief	-	-	X	
5	Oil seperators	-	X	-	
6	Oil temp regulator valves	-	-	X	
7	Oil flow meters	-	X	X	Samples
8	Oil quantity guages	-	X	X	

Airframe accessories

#	Item				Description
1	Braking systems	-	-	X	Samples or information on any new developments.
2	Cabin pressurizing	X	-	-	Including methods of protection against sudden decompression.
3	Cabin Pressure Regulators	X	-	-	Samples required
4	De-icer equipment	X	-	-	particularly information on research on: 1. Heated wing de-icing systems. 2. Cabin heating and de-frosting. 3. Windshield wipers, anti-icing sprays etc to maintain vision during icing and rain. 4. Methods of dealing with frost on parked aircraft. 5. Details of research establishments dealing with the above.
5	Energizers	-	-	X	Particularly AC generators - airborne power supplies.
6	Fire Detection	-	X	-	Details and samples.

#	Item				Notes
7	Fire Extinguishers	-	X	-	Any new types incl. extinguishing material.
8	Fluid & mechanical quick disconnects	1	-	X	
9	Generators	-	-	X	Other than main engine-driven, particularly AC systems.
10	Heating controls for flying clothing	-	-	X	
11	Heating systems for cabin	-	-	X	
12	Hyd pressure pumps	-	-	X	
13	Intl armor plating	-	-	X	
14	Instrument and cabin lights	-	-	X	
15	Lighting systems; external	-	-	X	
16	Oxygen cylinders	-	-	X	New or experimental types.
17	Oxy flow indicators	-	-	X	New or experimental types.
18	Oxy generators; airborne	-	-	X	Any type.
19	Oxy generators	-	1	X	Especially those using liquid oxygen.
20	Oxy pressure guages	-	-	X	New or experimental type
21	Oxy pressure regulators	-	-	X	
22	Oxy regs; constant flow	-	-	X	
23	Oxy regs; demand	1	-	X	
24	Oxy systems	-	-	X	
25	Oxy warning devices	1	-	X	Types indicating to pilot when any crew member stops drawing oxygen.
26	Relief tubes	-	-	X	
27	Safety belts etc	-	1	X	
	Airborne Communications Sets				
1	FuG-10	-	-	-	Obsolescent bomber radio
2	FuG-11	1	2	6	Liason type radio to replace FuG-10. May employ FM
3	FuG-15	1	2	6	New FM/AM command radio to replace FuG-16
4	FuG-16	-	-	-	
5	FuG-16Z, 16ZY etc		2	6	Command type radio.
6	FUG-17	-	-	4	Forerunner to FuG-16 series.
7	FuG-18	X	X	X	Details not available, may be similar to FuG-16 and 17.
8	New types of transmitters and recievers	2	4	6	Particularly evidence of decimeter or centimeter comms systems.
	Airborne Navigation				
1	FuB1-1	-	-	-	Blind landing equipment.
2	FuB1-2F	-	-	-	Blind landing equipment.
3	FuB1-2H	-	-	2	Blind landing equipment.
4	Peilgerat-6	-	-	2	Direction finding radio.
5	Peilgerat-7	-	2	4	Direction finding radio, may incorportae automatic radio-compass facilities.
	Countermeasures				
1	Jamming	X	X	-	Information on methods and/or specimens of equipment.
2	Anti-jamming	X	X	-	Information on methods and/or specimens of equipment.
	Ground Radio Equipment				
1	Fixed ground stations	-	X	-	Particularly VHF installations.
2	Mobile ground stations	-	X	-	Particularly VHF installations.
	Airborne Radar				
1	FuG-25-25A	-	-	6	IFF

#	Name				Description
2	FuG-101, 101A	-	-	6	Frequency modulated radio altimeter, later 101A desired
3	FuG-102	1	2	6	Pulsed radio altimeter
4	FuG-103	1	2	6	Radio altimeter. Reported similar to FuG--101A
5	FuG-120	1	2	6	Reported as a navigational device.
6	FuG-200	-	2	6	Anti-shipping airborne radar.
7	FuG-202	-	2	6	Aerial interception radar.
8	FuG-203A, 203D	-	2	6	Radio control
9	FuG-212	-	2	6	Modernised FuG-202. Transmitter and receiver in one case.
10	FuG-213	-	2	6	'Lichenstein S' anti shipping
11	FuG-214	-	2	6	'Lichenstein R' Decimeter tail warning. Appearance similar to FuG-202.
12	FuG-216	-	2	6	'Neptune Gerate'. Tail warning.
13	FuG-217	-	2	6	Reported development of FuG-216.
14	FuG-218	1	2	6	Reported development of FuG-216.
15	FuG-220	1	2	6	'Lichenstein SN_2' Development of FuG-202.
16	FuG-224 (Berlin)	1	2	6	German equivalent of British H_2S.
17	FuG-225 (Wobbelbeine)	1	2	6	IFF set.
18	FuG-226 (Neuling)	1	2	6	IFF set
19	FuG-350A (Naxos)	1	2	6	Centimetre D/F homing receiver (airborne) (various models exist)
20	FuG-351	1	2	any	Centimetre search receiver (ground)

Ground Radar

#	Name				Description
1	Benito		will be		Fighter control
2	Freya		handled		Anti-jamming devices only
3	Wurzburg		by special		Anti-jamming devices only
4	Mannheim		personnel		Similar to Wurzburg. Has enclosed cabin.
5	Jagdschloss		only		
6	New developments in mobile and fixed units using microwave radiation.				Any information or specimens recovered should be reported for examination by specialist personnel.

Gun laying Radar

#	Name				Description
1	All new types	X	-	-	Also items modified for special uses or equipped with anti-jamming devices.

Radar Countermeasures

#	Name				Description
1	All types of jamming and anti-jamming systems	X	-	-	All will be handled by specialist personnel -
2	Foil strips (chaff/window)	X	-	-	Suspected installation should be reported
3	VHF jammer - Protekt	X	-	-	
4	VHF jammer - Gereat	X	-	-	

Picture Resolving Devices (viewing devices)

#	Name				Description
1	Grob-Obi	1	2	6	Reported to use Nipkow disc as in early television systems.
2	Grob-Gerat	1	2	6	Development of Grob-Obi without Nipkow disc.
3	Thermal picture transformer	1	2	6	Any device for converting received infra-red or heat rays to visible light.

Aircraft indentification Devices that require infra-red source on aircraft and viewing device

#	Name				Description
1	Beaming devices	X	X	X	Usually used for formation flying
2	Portable devices	X	X	X	Used for pathfinder work for gliders or paratroops to assemble.
3	Searchlights	X	X	X	Used at coastpoints for ship detection.

Three Messerschmitt employees also joined the Whizzers: Karl Baur, the Chief Test Pilot of Experimental Aircraft who had replaced Dr Hermann Wurster in 1940, test pilot and former Luftwaffe pilot Ludwig 'Willie' Hoffman, (spelt 'Hofmann' in some sources) and Gerhard Caroli, who had been chief administrator of Messerschmitt's Department of Flight Testing. Test pilot Herman Kersting joined later.

Caroli, who had Italian ancestry, lived in a house across from the Messerschmitt office building in Augsberg and was well known to Karl Baur, for they had graduated together from Stuttgart Technical University in 1936. It was Caroli who, whilst at Baur's home, informed the former chief test pilot that Colonel Watson wanted them to assist in the retrieval of the Me.262. Their co-operation was to be voluntary, although Watson made it clear that he had other ways to make them co-operate. Baur did not wish to see all the work that had gone into the development of the Me.262 lost and so they co-operated, as Baur later recorded:

'Our testing centre at the airbase Lechfeld had been cut off from the home office of the Messerschmitt company in Oberammergau during the final three weeks of the war. Therefore, the management of our testing centre - including myself - were responsible for making a decision. We all knew that the development of the Me.262 was way ahead of any other aircraft development in the world at that time. After deep soul-searching by every individual, we met again and agreed that it was important to save this advanced technology at any price.

'The people of Germany had the right to know someday that during the darkest days of their existence something worthwhile was accomplished by their brilliant scientists and engineers. Therefore, we ignored Hitler's order to destroy all equipment and instead followed an order given by Speer, which stated that everything shall be paralysed only.

The Me.262s had been paralysed by removing the engine governors, which we had buried carefully - wrapped in oil paper. All we had to do was to recover those engines' controls and re-install them. The price we had to pay for our action was

Me262A-1A was, according to the records, surrendered by defecting Messerschmitt test pilot Hans Fay at Frankfurt/Rein-Main on 31 March 1945 after its maiden flight from Hessental. The location of this picture is unknown. *(USAAF)*

presented to us quickly, when certain groups of our countrymen called us traitors. We had to live with that weight on our shoulders for quite some time.'

Baur lived with his young family in Augsburg, catching a lift on a truck supplied by the Americans to and from the airfield at Lager Lechfeld. Hoffman, on the other hand, lived on the base, his wife and four young children living under Russian control in the east of the country. Hoffman had hoped that by working closely with Watson he could obtain American support to bring his family out of Russian-occupied Germany.

It is also interesting to note that it has been alleged and partially substantiated by declassified documents that the Whizzers recruited previously captured Luftwaffe personnel and pilots held as prisoners of war at Fort Bliss, Texas, to go into what would become the British, French and Soviet controlled areas after V-E Day, to fly out, hide, or otherwise remove all 'black listed' aircraft, secret weapons equipment and supporting documents to the U.S controlled areas some four months before Germany's surrender.

Arrival at Lechfeld...
On arrival Strobell noted considerable damage to the airfield: the runways had been carpet-bombed and few buildings were intact. Although there were a number of jet aircraft present, most were in a state of serious disrepair. It appeared that many had been intentionally destroyed by the retreating Germans, and the few that were left had fallen prey to souvenir-seeking soldiers and roving bands of displaced persons.

On reaching the Messerschmitt facilities – most of which were in ruins - he was pleased to see that a small group of Americans from the 54th Air Disarmament Squadron had preceded him onto the field. These men had arrived in the area a few weeks earlier with orders to preserve and safeguard as many Me.262s as possible.

The military government had already located a number of German nationals living in the area who had worked on the Me.262 programme. These technicians had been placed under contract as civilian employees to assist in the 54th ADS effort.

'Watson's Whizzers' team - from left to right: Holt, Haynes, Anspach, Watson, Dahlstrom, Hollis, Strobel, Maxwell.

The Germans were justifiably proud of the jet, and were prepared to give the Americans their full cooperation, already having succeeded in preparing several machines for flight.

A week prior to Strobell's arrival, the last of eight airworthy 262s had been test flown, and two more were awaiting engines. The 54th ADS men were quick to make their mark upon the project by painting names on the left side of each of these aircraft - the right side of each jet bore their unofficial squadron name, born of their constant squabbling: *The Feudin' 54th*.

Although none of the promised pilots were yet on hand, three or four of the crew chiefs assigned to Colonel Watson's project had arrived. Strobell found them in a bombed out hangar with rifles at the ready, awaiting their instructions. Although the language barrier had prevented them from communicating with the German crew, they reported that so far they had not encountered any problems.

While the men decided to stay in the hangar, Strobell spent the first few nights on the second floor of a bombed out administration building. Still wary of his surroundings, he kept his .45 nearby and laid a string of cans across the stairwell as a precaution.

The work of the ADS was done, and they left the field to Strobell and his mechanics on 2 June. A day later, two more pilots arrived: Lieutenants Ken Holt and Roy Brown. They were soon followed by Lieutenant Bob Anspach and the rest of the men.

It was not long before the entire team was assembled: six pilots, ten crew chiefs and some two dozen German nationals. Watson was away tending to other matters and was rarely present, but the men had a clear understanding of their mission, and set to work immediately.

Among their civilian employees were two English-speaking Messerschmitt test pilots: Ludwig Hoffman and Karl Baur. Both were cooperative and professional, though the men took an immediate liking to the more good-natured Hofmann, whom they began calling 'Willie'.

A personal friend of Charles Lindbergh and a legendary aviator with a reputation throughout Germany, Hoffman had flown virtually every type of aircraft, including, allegedly, the Bachem Ba.349 'Natter' rocket-propelled interceptor. Few knew the Me.262 better than Hofmann, and he did his best to convey to the young Americans how to stay out of trouble in the jet.

The men learned from Baur that one of the aircraft - now named '*Beverly Ann*' - had been surrendered intact near Munich, and had been flown into Lechfeld prior to their arrival. Another was flown to Lechfeld directly and surrendered on VE Day. There was also an original factory trainer on site

ME262A-1A 'V083' seen at Lechfeld, but now named '*Wilma Jeanne*'...

...before being renamed one more time as *'Happy Hunter II'* as seen here at Melun. The machine had been flown from Lechfeld via St. Dizier to Melun on 10 June 1945, but later crashed on route to Cherbourg. The nose contours of this machine was very distinctive. *(USAAF)*

that still remained in a flyable state. Otherwise, they were told that the majority of the team's aircraft had been built from an odd collection of engines, various nose sections, landing gear components and parts scavenged from wrecks.

Freiburger explains more about working on putting together complete 262 airframes. *'We identified the aircraft by the numbers painted on the fuselage or by the names that we had christened them with. For example, the aircraft that I named after my wife, the 'Wilma Jeanne', had a number on it, V083. This aircraft had a 50m cannon and carried 22 x 50mm cannon shells in a chain link belt in the nose. The two-seat trainer I named after my sister-in-law, 'Vera'. The other master sergeant in my squadron, Preston, had named one after his wife, and called it 'Connie The Sharp Article' - but when the American pilots came in, they changed some of the names on the aircraft. Col Watson changed the 'Wilma Jeanne' to 'Happy Hunter II' after his son, and so forth.*

'The two-seater trainer, an Me.262B, was changed from 'Vera' to 'Willie', after Hoffman. This aircraft had the least amount of damage and was readily put in flying condition with the minimum amount of work. My operations officer, Capt Ward, who would occasionally come out to check on us, went for a ride in this aircraft. Consequently, he became the first American to ride in a German jet.'

It was decided that the ATI team should bring each of these aircraft back in turn into the hangar for a more detailed inspection. The earlier ADS effort had been conducted in understandable haste, and under minimal American supervision. No one could completely rule out thoughts of possible sabotage, and these additional inspections provided the crew chiefs with a very good opportunity to learn about the systems of the unusual aircraft.

As the work progressed, the mechanics found increasingly innovative ways of communicating with their German counterparts, and activities in the main hangar were in full swing. Inspecting, repairing and rebuilding was accomplished as necessary, with either Hoffman or Baur conducting a new test flight as each jet came out of the hangar.

Each crew chief was assigned a specialty area, and quickly became proficient in his area of expertise, while the pilots rehearsed engine starts and reviewed performance characteristics on a damaged Me.262 that had been tethered to the ground. In the space of just over a week, all ten machines were refitted, checked out and ready to fly.

On 8 May 1945, Eugene Freiburger accepted the surrender of Lt Fritz Muller, the pilot of Me.262A-1a WNr 500491, which was rapidly christened *'Dennis'* after Eugene Freiburger's son. In this aircraft, Ofw Heinz Arnold had earlier achieved at least seven combat victories.

With the aircraft emerging from the hangar and the pilots on hand, there was only one small detail which remained: none of the Americans had yet flown the jets. Strobell was determined to get at least one flight under his belt so that he could give the others some idea of what to expect. Hofmann and Baur were superb test pilots, but they were not professional instructors. Strobell knew that it would be better for everyone involved if he got the first solo out of the way – as long as he survived it!

He later recalled the details of this flight: *'The first Me.262 restored was an Me.262A-1. I could be wrong, but am almost sure it was the one named Beverly Ann (this was my cousin's name, and I wondered how her name came to be on the airplane). This was the Me.262 that Baur flew into Lechfeld on 16 May from Munchen-Reim near Frankfurt. It was found in fully operational condition, requiring the least amount of hangar / crew chief attention.*

The first few days of June, this airplane was undergoing a complete check, and I was aware that it would be the first out of the hangar. At the time, I was still a bit leery of the entire Messerschmitt crew as a whole, feeling that it took only one bad apple in the lot to spoil our plans. So I went to the shop superintendent, Mr. Caroli, and told him that Baur would make the test flight. My thinking was that if this was generally known by the crew there would likely not be an attempt to sabotage it ... it's called 'finesse'.

When the airplane rolled out of the hangar it was refueled with a limited load. I asked Baur to make the test flight, which he did. He was up about fifteen minutes and landed. When he touched down on the runway, I was sitting at the approach end of the runway in a Jeep. An enlisted man was driving, and Ken Holt and Bob Anspach were along as passengers. When Baur touched down we were at full speed, racing down the runway to catch him. We met him just as he was about to U-turn to taxi back. I asked him to step out of the jet, and he did. I climbed in and taxied back to the hangar where we refueled it with a full load. Then I taxied out to the runway.

My first solo flight in the Me.262 started with a pilot error on takeoff. Somewhere in the back of my mind I got the impression that swept wings required a higher angle of attitude on takeoff. It must have come from watching Baur make his takeoff. About halfway down the runway, all was going well, except that I noticed that I was gaining flight speed slowly, if at all! Everything was roaring along just fine, except the airspeed was not up to takeoff, and didn't appear to be increasing as rapidly as expected. At this point I lowered the nose and put the nose wheel on runway. I was doing something like 70 or 80 miles per hour, and up came the airspeed ... I found myself at the end of the runway, and I simply hauled it off of

The Henschel Hs 293 was an anti-ship guided missile: a radio-controlled glide bomb with a rocket engine slung underneath it. It was designed by Herbert A. Wagner.

the ground, feeling that I had used all 6,000 feet of a 5,000 foot runway. One is not likely to forget such an adventure, and I still think about that rough trip down that runway as we watched the end approach - both mine and the runway's.

The next surprise came when I was climbing out, reaching for altitude. The wing slats started blinking in and out. I thought that they would stay out or snap shut closed. They didn't. They would close momentarily with a bit of air turbulence and then open again. This continued for a brief period, like a minute or so, until the airspeed increased and the air pressure kept them closed.

The next thing I noticed was the speed. Raw speed, exhilarating speed. Smooth speed. Unbelievable speed. It seemed effortless. My flight was held to low altitude, so I had the ground as a reference. This was something I had never experienced in the P-47 Thunderbolt, and it was impressive.

But ... with the speed came another surprise. Air turbulence at the cruising speed of the Me.262 affects the airplane in ways that I had never felt before. An updraft became a 'butt thumper,' more like a jolt ... it was the same with a downdraft, so that on a hot summer day, at low altitude, you literally bumped and thumped *your way across the country. I thought those toe straps on the rudder pedals were humorous until I found out why they were there ... those sharp bumps would lift your feet off of the pedals.*

When it came time to return to Lechfeld to make a landing, I committed my second pilot error. I made a normal 'P-47 approach' to the landing by entering the downwind leg. I was planning for a quick left turn onto base and then final, but I never got out of the downwind leg! Normally, with a Thunderbolt, you would pull

The 'Category One' List contained bombs, missiles and guns.

Item No.	Description	I	II	III	Remarks
	Aerial Torpedoes				
1		1	2	10	Reported to be a glider attachment to allow the torpedo to be released from the aircraft at a distance.
	Flying Bombs				
1	V 1 (FZG-76)	1	2	4	Only larger versions of V 1, or version with different propulsion system desired. Also pilot operated types.
	Guided Missiles				
1	Ensian	1	2	10	Rocket driven small aeroplane similar to Me.163 - remotely controlled reported. Built at Augsburg.
2	Hs-117	1	2	10	Reported rocket propelled, radio controlled, ground launched against bomber formations. 10 ft long, 7.5 ft span. Sharply tapered wings.
3	He-293	-	-	4	Early glide bomb. Radio controlled.
4	Hs-298	1	2	4	Air launched, winged missile, controlled by a cable or radio.
5	V-2	X	X	X	Everything in connection with V-2, including launching and servicing equipment should be guarded and reported at once.
6	Wasssserfall	1	2	4	Reported 25 ft long, similar to V-2. Launched from hole in ground.
7	X-4	1	2	4	Reported small rocket bomb, launched from fighters. Small wings, controlled by cable; proximity fuze.
8	X-7	1	2	4	Reported ground launched rocket, remotely controlled.
9	BP-20	1	2	10	Rocket-interceptor. Wood build. Me-163 engine. Span about 13 feet.
10	Natter	1	2	6	Reported rocket-driven aircraft - pilot operated. Driven by 24 - 37mm rocket shells.
11	Igel	1	2	6	Reported as Natter with 12-15 75mm rocket shells.
12	Rheinsochter	1	2	10	Reported rocket-driven remote controlled shell - 2 types a = speed of sound. b = slower speed.
13	Schmetterling	1	2	10	Rocket driven shell - not remote controlled.
	Ammunition & Ammunition Accessories				
1	Ammo for Mk5, 55, 411	X	X	X	Adequate supplies for firing tests.
2	Hollow charge for Mk.108	X	X	X	
3	High velocity ammo for MG 131, 153	X	X	X	
4	Ammo boosters	-	-	X	
5	Ammo feeds/mags	-	-	X	
6	boxes, chutes, cases, links	-	-	X	
	Bombs and Bombing Accessorories.				
1	Armour Piercing	-	-	X	Samples of all types

#	Item				Notes
2	Chemical bombs	-	-	X	Samples of all types
3	Fragmentation	-	-	X	Samples of all types
4	General Purpose	-	-	X	Samples of all types
5	High Explosive	-	-	X	Samples of all types
6	Incendiary	-	-	X	Samples of all types
7	Semi-Armour piercing	-	-	X	Samples of all types
8	Bomb Disposal equipment	-	X	-	Any used on American bombs and long delay fuzes.
9	Bomb Bay systems	-	-	X	
10	Bomb handling equipment	-	-	X	Cranes, trailers, lift trucks etc.
11	Bomb release mechanisms	-	-	X	
12	Bombsight calibration stands	-	X	-	

Bombsights

#	Item				Notes
1	Lofte-7D	-	-	4	Standard bombsight
2	Lofte-7H	1	2	4	Development of Lofte-7D. High altitude sight, used with 3-axis autopilot.
3	Lofte-7K	-	2	4	One recovered
4	Navi-	1	2	4	Details not available.
5	Doranth Platte	1	2	4	A fitment to the BZA sighting head.
6	New Revi (gyroscopic)	1	2	4	A gunsight also used as a bombsight.
7	Tief-Schlender-Anlage (TSA)	-	2	4	Low altitude 'slinger' sight. Development of the BZA sight. TSA-2 recovered.
8	Fuses	-	-	X	Samples of various types.

Chemical Warfare

#	Item				Notes
1	Chemical Warefare bombs	-	-	X	
2	Chemical Warfar agents	-	-	X	
3	Defensive masks	-	X	-	

Guns, Cannons and accessories

Ammunition to accompany guns. Several hundred rounds to accompany the guns are requested.

#	Item				Notes
1	Recoiless guns of all types	X	X	X	
2	Any cannon 40mm or over	X	X	X	Reported experimental types
3	Bk-5	-	2	6	5cm aircraft cannon
4	Mg-153	1	2	6	Believed modified Mg-131, barrel for high velocity ammo.
5	Mk-55	1	2	6	5cm aircraft cannon
6	Mk.103	1	2	6	3cm long barrel cannon
7	Mk-107	1	2	6	3cm cannon - believed similar to Mk.108
8	Mk-108	1	2	6	3cm short barrel cannon
9	Mk-112	1	2	6	No details available
10	Mk-411	1	2	6	Believed to be 28mm cannon
11	Gun chargers	X	X	-	Samples of types found
12	Gun Mounts; flexible, fixed	-	X	X	20mm gun mounts particularly (power driven and manual controlled). Particularly for suspending guns under wings or under belly.

Gunsights

#	Item				Notes
1	Zeiss Gyro Gunsight	X	-	-	Carl Zeiss of Jena reported to be developing gyro gunsight; said to have been destroyed.
2	Any Gyro gunsight	1	2	4	
3	Fire control computors	1	2	4	Particularly remote controlled.
4	Fire control stations	X	X	X	Details and equipment

Turrets

#	Item				Notes
1	FDL-B 131/2A	-	-	3	Any new developments
2	FDL-B 131/2F	-	-	3	Any new developments
3	FDL-B 131/2K	-	-	3	Any new developments
4	FDL-B 131/2B	-	-	3	Any new developments

The Ruhrstahl Ru 344 X-4 was a wire guided air-to-air missile designed by Germany during World War Two. It did not see operational service and thus was not proven in combat. However, the X-4 was the basis for the development of experimental, ground-launched anti-tank missiles that became the basis for considerable post-war work around the world. *(USAAF)*

the throttle back on the downwind leg, drop the gear, and make a U-turn back toward the runway controlling speed with the throttle while descending to touchdown.

In the Me.262, I pulled the throttle back and nothing happened. I mean that there was no apparent reaction from the airplane. It simply continued to fly at the same speed, and I recall thinking that I had discovered 'perpetual speed.' By the time I figured out that I wasn't on a normal jet approach, I was five miles beyond the airfield, and still headed outbound at high speed. The airport had long since disappeared from sight! We had been cautioned not to reduce the turbine below 6000 RPM in the traffic pattern, but it seemed that this only encouraged the jet to continue to fly at cruising speed.

I finally turned back toward the airport and again entered the downwind leg at 500 feet. But this time I had figured things out, or so I thought. I pulled the RPM back to 6000 and pointed the nose up in a climb attitude. The airspeed dropped to 250 MPH, at which point the landing gear could be lowered safely. I managed all of this with my head inside the cockpit, so you can imagine my surprise when I discovered I was at 2,500 feet and again several miles from the airport. I continued around for a third and final approach, and landed without incident.

After landing, I taxied back to the hangar area and climbed out of the jet. I related all of these experiences to the other pilots, and they capped my day with a real class act event: Ken Holt and Bob Anspach walked over to me, and without hesitation removed my Army Air Corps insignia. They broke off the propellers and stuck the wings back on my collar while advising me that I no longer needed the propellers, since I was now a jet pilot. It was perfectly timed and precisely appropriate for the moment. It remains an indelible event in my memory.

This was to become something of a unit tradition. After each man completed his first flight in one of the jets, the propellers would be removed from his AAF insignia. For the rest of the mission, all of the men (except for Watson, who was always in and out of various headquarters) wore the non-regulation collar brass everywhere they went.

Names and numbers

For both administrative and technical reasons, after a thorough-as-possible servicing, the aircraft were each given triple digit identity numbers starting with '000', usually painted on the nosewheel doors. Once '999' was reached a new series of numbers started with '101'. Most of the aircraft were also renamed.

Colonel Watson - who his crews had nicknamed 'Grey Wolf' - used a somewhat dilapidated Douglas C-47 which he called 'his office', for transporting spare Jumo 004 turbojets and other equipment around. While at Lager Lechfeld Watson held discussions with Professor Willy Messerschmitt, whose home was only a short distance away, and Messerschmitt made it clear he was delighted that his work on the Me.262s and other aircraft designs would not be lost. Watson also saw it as a major priority that skilled German personnel were sent to the USA: *"We couldn't take them all back, so we concentrated on those who had been involved in the technical programme, to get hold of the scientists, documents, and wind-tunnel test data and anything else that was relevant. It was a major priority as far as I was concerned to bring back the tehnicians and scientists."*

The remainder of the pilots were given brief orientation flights in the two seater on the 9 June. Most of the records pertaining to the equipment found on the field appear to have been destroyed or otherwise lost. Since so little was known about the history of the engines and similarly critical components, this was the only formal flight training they dared to undertake. Aside from what they had managed to pick up while operating the ground trainer or practicing blind cockpit drills, the pilots were basically on their own. Most of the men would have to experience their first solo flights in the Me.262 on the extended cross-country flight to France.

Watson warned his pilots that the approach and landing procedure for a jet aircraft was considerably different to the normal approach pattern of a piston-engined aircraft. He advised them that the Me.262 would require a longer approach to the airfield, and that undershooting and overshooting needed to be rapidly anticipated in view of lack of responsiveness to the throttle.

Initially it had been planned to fly the Me.262s, together with other aircraft collected by Watson's team, over to England to meet the British auxiliary carrier HMS *Reaper*, which was berthed at Liverpool docks. However, with only three weeks' notice, Col Watson learned that *Reaper* would call in at Cherbourg on 1 July to collect the aircraft, on the way to New York. Watson's task was to get the Me.262s to Cherbourg/Querqueville by this date. There were no aerial maps of the route, so the journey to Cherbourg was planned from the Michelin road maps hanging in the office at the Lechfeld airfield. It was originally intended to make one stop at Melun, some 25 miles south of Paris.

The team of fledgling jet pilots quickly experienced considerable main undercarriage brake trouble, which was attributed to excessive use during the first three landings they made. Watson remedied this error by providing his pilots with additional training. He recalled; *"The Me.262 was designed to*

262A-1a was coded 'Yellow 7' of IV/JG7 before recieving the name *'Dennis'* and later *'Ginny H'* after Lt James Holt's girlfriend. It is seen here at Lechfeld with wire guards over the engine intakes. *(USAAF)*

take off from dirt strips, and we found that when we eventually reached Melun, on the way to Cherbourg, it was mostly a dirt strip. We realised, just by looking at it, that the undercarriage nosewheel was a weakness on the Me.262, and we trained our pilots to operate the nose-wheel with gentle care."

The ferry operation to get the aircraft out of Germany commenced on 10 June. The machines were lined up on the taxiway in a single file, and a final check was conducted early that morning. Captain Hillis had prepared an operations plan which called for the first takeoff at around 0930.

For safety reasons, each takeoff time was delayed by ten minutes in order to provide the ground crews adequate time to clear the runway at the destination airfield in the event of any mishap. This was a legitimate precaution: none of the men had any real experience with landing the jet, and the landing gear itself was known to be somewhat failure-prone.

As the team consisted of only eight American pilots, two cockpits would have gone unfilled without the aid of the Germans. Both Hofmann and Baur were retained to ferry aircraft, with Hofmann taking the two-seater trainer (later known as *Willie*) and Baur flying a standard fighter model (later known as *Jabo Bait*).

Earlier in the week, each of the Americans had been given responsibility for a specific jet, and this was carried over when the flight assignments were made (it also was to take on a greater significance in Melun, when new markings were applied to each machine).

Lieutenant Roy W. Brown recalled his first jet flight: *'The Me.262 was smooth, quiet, and very responsive to the controls compared to the P-47 I had been flying for about a year. I had also flown a P-40 in the States, and the Me.262 was even better than that.*

The plane was easy -- and a pleasure -- to fly. Because of its high speed, I found myself going through my maps quickly to keep pace with the distance covered over the ground.

I glanced at the engines periodically. The engine tail pipe had a moveable cone, reducing the cross-sectional area of the exhaust gases when the engine was advanced to full power. The cone automatically moved rearwards as the RPM increased and would be extended at full power.

Another feature was the moveable leading edge of the wing. This moved forward automatically when the air speed dropped below a set speed forming a slot through which the air could flow to the top of the wing. This helped maintain improved air flow over the wing, reducing stall and landing speeds. After take-off the leading edge slid back automatically as the air speed increased.

Surprisingly little difficulty had been experienced with the Jumo 004s on the flight from Lager Lechfeld. In a report dated 13 June, Colonel Watson detailed operating the Me.262:
Object: To report on the Jumo 004 engine operating procedure and flight characteristics of the Me.262 airplane obtained through preliminary flight tests and

The two-seater 262A 555 'Vera' with the number 35 on the side of its fuselage is seen on 10 June 1945 awaiting delivery from Lechfeld to Melun. *(USAAF)*

Connie The Sharp Article or is it *Connie My Sharp Article...* whatever the name, this Me.262A-1a/U3 was a photo-reconnaissance variant that was later renamed 'Pick II'

ferrying of the aircraft.

The Me.262 airplane with Jumo 004 Bl engines installed was given a preliminary test flight at Lechfeld, Germany, and subsequently ferried to Melun, (A-55), France. It was originally intended to fly the airplane direct. However, due to failure of the auxiliary fuel pump in the 600 litre auxiliary tank, a stop-over for fuel was necessary at St Dizier (A-64), France.

The first flight consisted of normal take-off and climb at 450 kmh to 9000 ft with full throttle. During the climb, uniform fuel pressures were maintained at 40 ATU's at approximately 1800 ft per minute rate of climb. A 'rolling-moment' to the left was experienced which could be readily corrected by increasing the power of the left engine to approximately 42 ATUs or by applying 180 degrees clockwise turn of the rudder trim with equal engine fuel pressures. At 9000 ft altitude with full power, 830 kmh maximum speed was obtained for approximately one minute.

During a diving turn to the left with rudder trim at zero and equal fuel pressure of 20 (ATUs), a 'rolling moment' to the left occurred which, if not corrected by use of left engine throttle, would have resulted in a slow roll to the left. It was apparent on this flight that some minor jockeying of the throttle is necessary for controlling lateral instability. At no time did engine exhaust temperatures exceed 650° Centigrade. With wheels extended and full flaps and engine throttled to 5000 rpm at 5000 ft altitude, the airplane has normal controllable stall characteristics at approximately 205 kmh with both main tanks containing 700 litres of fuel. The slots at the leading-edge of the wing extend at speeds varying between 400 and 430 kmh, depending upon the angle of attack of the airplane.

Extensive flight tests were not performed as the purpose of these flights was to secure range data and flight characteristic data for ferrying additional jets, and training additional pilots.

During the ferrying of the first airplane from Lechfeld, Germany, to Melun, France, the fuel consumption at an average altitude of 3000 ft at 600 kmh and 18 ATUs fuel pressure was approximately 2700 liters per hour. During ferrying of the remaining nine airplanes from Lechfeld to Melun, the fuel consumption averaged 2250 litres at an altitude averaging 10,000 ft at 615 kph at 18 ATUs fuel pressure.

During the preliminary flight tests and the ferrying of the Me.262s, very little difficulty was experienced with the Jumo 004 engine. Two failures of the two-cylinder starter motor clutch were experienced, necessitating changing of these starter units, consuming approximately 30 minutes. One exhaust cone was found to be slightly cracked. Considerable main gear brake trouble was had due to overly excessive use during the first three landings of the Me.262 by newly trained pilots. It is anticipated that additional training will overcome these excessive difficulties in due course.

Fuelling was less of a problem than was expected - at Lechfeld the 262s were serviced with J-2 fuel. At St Dizier however, the aircraft were filled with standard diesel fuel obtained from the Engineer Corps!

The weather was good and the field at Melun was easily visible. By nightfall, the entire team had completed their ferry flight to Melun without

An inspection at Melun. The day General Spaatz inspected Watson's Whizzers was dreary and wet, as this picture of the flightline shows. The name on the 262 furthest from the camera is not readable, but is thought to be *Screamin Meamie*, then there is *Ginny H*, and closest to the camera is *Jabo Bait*. (National Air & Space Museum)

incident. Ten of Germany's most advanced jets were safely under guard at the French airfield at Melun, never to return to their homeland. This was to be an intermediate stop along the way to a port at Cherbourg.

Lieutenant William V. Haynes joined the Whizzers at Melun, and went for a quick checkout ride with 'Willie' Hoffman in another recently obtained trainer. He was the last addition to the team, raising the total to nine men, including Watson.

On 11 June, Colonel Watson sent a memo to Colonel Huntington D 'Ting' Sheldon, Assistant Chief of Staff A-2:

'For your information the project of ferrying 10 Me-262s from Lechfeld, Germany to A-55 (Melun, France) has been completed. These airplanes are being serviced and maintained at A-55 and will be flown to A-23 Cherbourg for preparation for overseas shipment prior to loading on Navy flat-top.

This project consisted of the following steps:
a. Assembling the Me-262s by former Messerschmitt German employees.
b. Training 10 airplane crew chiefs on airplane and jet engine maintenance.
c. Determining maintenance history of the Me-262 airplane and engine and securing five truckloads of spare parts, maintenance tools, and special purpose maintenance equipment. These five truckloads of equipment will arrive at A-55 today, Monday 11 June.
d. Training 6 American pilots on the operation and performance of the jet airplane as well as giving each pilot a flight test in a two seater Me-262 trainer.
e. Establishing refuelling points and emergency landing fields along the route from Lechfeld to A-55.
f. Personally flying the first Me-262 from Lechfeld along the proposed route to Melun to definitely establish the feasibility of the project and to secure range data and performance data on the airplane to enable training of American pilots.

The ferrying project started at 0925, Sunday 10 June and was completed at

262A-1A *"Screamin Meamie"* - formerly named *'Beverley Anne'* is photographed at Melun in June 1945. The aircraft was allocated to Lt. Robert Strobell.

Me.262A 222 named *"Lady Jess IV"*, allocated to Captain Kenneth Dahlstrom. The large 'teardrop' fairings on either side of the nose forward of the cockpit was to accomodate the film magazines for the two Rb50/30 cameras installed alongside a single forward-firing 30mm cannon.

1915 without any major incident. Minor maintenance propblems occured which served to raise the experience level for continued flight and performance testing of these airplanes.

At A-55 Melun there is presently located the following:
a. Six trained pilots and three additional pilots are undergoing training.
b. Ten trained American crew chiefs who are capable of performing first and second echelon maintenance on these airplanes and in a few days of additional training will be capable of supervising third and fourth echelon maintenance.
c. Five 2.5 ton truck-loads of spare parts, maintenance tools, special purpose maintenance equipment including one spare Jumo 004 engine.
d. 2 of Messerschmitt's best test pilots.
e. 1 Jumo engine specialist.
f. 1 Messerschmitt technical advisor.
g. 1 Me-262 Messerschmitt technical inspector.
h. 9 Messerschmitt aircraft mechanics who are assisting American crew-chiefs in performing maintenance.

It is strongly recommended that the Me-262 airplanes along with the 10 trained crew-chiefs and pilots are shipped and transferred to Wright Field or other research centers in the United States as a unit in order to accomplish performance and flight testing and comparative analysis with similar American equipment. From the experience gained while carrying out this project by the undersigned, it is believed that the research centers in the Zone of the Interior would profit considerably should this project be carried through as outlined above'.

In addition to checking various locations for other aircraft and parts, the men readied themselves for a command review by General Carl Spaatz while at Melun. This had been postponed several times, but eventually took place on 27 June. The plan called for lining the aircraft up on the ramp with a pilot and crew chief posted at each, followed by a brief aerial demonstration.

Spaatz and his team arrived at mid-morning, inspecting and asking questions of the men. He was visibly impressed by what they had accomplished, and spent considerable time examining the aircraft.

Following the static review on the ramp, Strobell, Holt and Hillis moved their jets into position for the flight demonstration and took off. Anspach and Brown were standing by, ready to launch their machines if needed.

The runway was damp from a light misting that had come and gone for most of the morning. This made for an impressive display as the jet blast kicked up long rooster tails in their wake. Hillis' jet experienced a problem while retracting the landing gear, but Strobell and Holt commenced a series of high speed, low-level passes over the runway. Strobell then initiated a series of rolls over the field.

The two-seat trainer 262A 555 was renamed *'Willie'* in honour of Willie Hoffman who checked out many USAAF pilots on the type. It is seen here at Melun awaiting review by Colonel Carl Spaatz. *(USAAF)*

With the Top Brass duly impressed and the formalities behind them, the team entered the next and final phase of their mission: ferrying the jets to the port city of Cherbourg. There arrangements had been made with the military port authorities to load the aircraft onto a ship for the trip home, as this cable from Brigadier General George C McDonald of the Exploitation Division shows. The cable also demonstrates that there was more than just Watson's Whizzers to consider:

Now available for priority shipment to Zone of Interior are fifty flyable aircraft, jet, rocket, and other types together with essential spare parts such as replacement jet engines required as Category One specimens for research design development of most recent experimental enemy equipment. Such equipment necessary to insure the technical superiority of our own equipment.

Pooling resources of captured priority enemy materiel with United States Naval Technical Mission in Europe message dispatched to Chief of Naval Operations indicating disassembly and reassembly of flyable aircraft in the United States after normal shipment would result in damage to aircraft and loss of components thus making it advisable if not mandatory that an aircraft carrier be made available until 11 July for approximately 11 days for shipment of above priority research German aircraft. Request co-ordination with Naval Chief of Operations to insure availablility of carrier.

With the war still raging in the Pacific, the only suitable vessel available was the British escort carrier HMS *Reaper*, first commissioned for the US Navy as the USS *Winjah*. The Royal Navy agreed to support the project, and the pilots made plans for the final leg of their journey.

The flights to Cherbourg - in particular from the small airfield at Querqueville, also known as ALG A-23C Querqueville - were conducted singly between the 30 June and 6 July. The earlier aircraft and pilot assignments were generally repeated with some exceptions, and the only timeline of importance was in making the rendezvous with HMS *Reaper*.

The sole catastrophic turbo failure experienced by the unit led to the loss of Me.262A-la/U4 WNr 170083 on the flight from Melun to Cherburg. This was V083. the prototype 50 mm cannon-armed Me.262 that had been flown to Melun by Watson - who named it *Happy Hunter II* after his son. The mission turned into a brush with death along the way, one of the engines began shedding turbine blades, and the resultant vibration caused a tailplane malfunction that placed the jet into an uncontrollable dive. But the pilot, 'Willie' Hoffman, managed to bale out at low altitude. French farm workers discovered him with a badly broken leg. This highly significant aircraft was one of only two test vehicles built by Messerschmitt to test the long MK2 14 cannon. Watson remarked that *'The recoil system was so good that you could hardly feel it in the cockpit'*.

In broken French he asked them to contact the nearest American military station. The French were all for lynching the German pilot, but after hours of pleading a US Army officer arrived and, after the officer was shown a pass

Me.262B '999' *Ole Fruit Cake* at Melun.

signed by Col Watson, Hofmann was moved to the American Military Hospital in Paris, where Watson subsequently visited him.

At Melun, Messerschmitt's experimental chief test pilot Karl Baur accompanied Col Watson and Capt Fred Hillis to collect three Arado Ar 234s from Sola, a former Luftwaffe base in Stavanger, on the south-west coast of Norway. Roy Brown recalls one of these missions: *'We had a P-47 assigned to our group, and I remember flying in it to Schleswig, Germany and then on to Grove, Denmark, checking on German planes. The airfields at both places were under British control, and the British were very helpful. At Schleswig we found a second two-seater trainer Me.262 and a night-fighter version. At Grove there were two Arado Ar 234s which had come from Norway. The four planes were flown to Melun to add to the collection. Later, two more Ar 234s were flown from Norway by way of Grove.'*

After flying the Ar 234s back to Melun, Baur left the Whizzers, who, from that point, continued their journey to the USA via Cherbourg without any German personnel. For the Germans, who had been promised a trip over to the USA, the next few months were fraught with anticipation.

The Whizzers flew the Me.262s and other aircraft including the Arados from Lechfeld to St. Dizier to Melun and then to Cherbourg.

Roy Brown's flight in 444 on 5 July was characteristically uneventful, but Anspach's experience in 333 was somewhat less straightforward. He departed Melun at 0930 hours on 30 June - intended destination Cherbourg. Weather en route was low broken to scattered clouds with a visibility of 10 miles or better. Because there were no air to ground communications, the flight was intended to be made under the cloudbase. About 30 minutes out Anspach decided to top the cloud - which was very thin - with the intent of dropping down through it in ten minutes. This should have brought him down east of Cherbourg, but on breaking through the cloud again on descent, he found himself over water with no land in sight.

He made a turn to a heading of 90° knowing that it would return him to land in approximately three to five minutes, and the coast should appear. A fuel check showed he could make it but with little to spare.

'At this time an island came into view (the island of Jersey). *I could see a landing field -- it was a very short grass strip. Being low on fuel I made the decision to land. I made a left-hand approach turn, and on final, I reduced speed to just above stalling with gear and flaps to full-down position.*

There was a church steeple at the approach to the runway. Witnesses stated that as I came over the church, the steeple tip went between my landing gear. I touched down within the first 200 feet of the runway and immediately started gentle application of the brakes. I experienced very little difficulty getting the plane stopped before reaching the end. I attributed this to the grass runway slowing me down. Had it been paved, I would have gone off the end where there was a considerable drop. Later, I was told the runway was 3800 feet in length with a sheer drop at the departure end of 250 feet down to the sea!

It took about 24 hours for my whereabouts to be known. Messages were communicated from the island to London (fighter operations) and then to Melun to Colonel Watson. Once my location was known a C-47 with several of the maintenance crew was dispatched to ascertain what had to be done. A decision had to be made on whether to fly the jet out or have it dismantled and barged to Cherbourg. If we did the latter it was then questionable if we could meet the Reaper's departure schedule.'

At the time of this incident, Jersey Airport was operated by a detachment of the RAF's 160 Staging Post, which had it's headquarters at Guernsey and provided refuelling facilities for service aircraft, mainly Dakotas and Ansons flying from Croydon. What the detachment's Commanding Officer, Flt Lt Aranson thought when he and the air traffic controller saw the Me.262 on approach can only be imagined!

The length of the runway at Jersey was of great concern - particularly with the sheer drop at the end. Barrels of jet fuel had been brought in on the C-47 in anticipation of this event, but even so, there was only just enough fuel to make the flight to Cherbourg.

The jet was towed to a position allowing for a maximum take-off run. The takeoff was routine and, amazingly, Anspach did not have to use all the runway. He climbed to 5,000 feet and proceeded to Cherbourg where the landing was uneventful.

Anspach was in for another wild ride a week later. This time, he was ferrying one of the two-seaters (#555) to Cherbourg.

On 6 July he departed Melun at 1000 hours and the trip was uneventful until the landing approach was initiated. Upon lowering the landing gear, Anspach received indication of the main gear extending but no panel light that the nose gear was down. He activated the emergency gear-down switch, which was a compressed air cylinder, but still did not receive a gear-down indication.

Earlier he had received a green light from the tower to land so continued on the approach expecting a red light if the nose gear was not fully extended. He thought the gear was down and that the down-indicator was unreliable. He touched down normally on the main gear, holding the nose off the ground as long as possible. After rolling a third of the length of the runway, he slowly lowered the nose and discovered that he did not have gear extension.

The aircraft slid for 800 to 1000 feet straight ahead on the nose section, engine nacelles and main gear before stopping with very little damage inflicted to the aircraft.

All that was required to make it flyable was to replace the nose section and

A Focke-Achgelis Fa.223 E 'Drache' in the charge of the Americans.

Four views of the Focke-Achgelis Fa.223 E 'Drache' twin-rotor helicopter at Airborne Forces Experimental Establishment, Beaulieu after painting in RAF markings. This machine was one of a number of German helicopters evaluated by the British and started its trip to the UK in May 1945 when it was surrendered to American forces at Ainring. A number of helicopters were re-painted in US markings and flown to Munich, and then to the US Air Technical Intelligence Unit at Nellingen, near Stuttgart. On 15 June it was flown to Villacoublay near Paris, en route to Cherbourg for shipment to the USA, but due to lack of shipping space, it returned to Villacoublay on 20 June and soon after was released to the British. It was not until 4 September that it was flown to Le Havre by test pilot Hans-Helmut Gersenhauer, departing again on 6 September for Abbeville where he re-fuelled before taking off for Lympne in Engand, this was the first helicopter to cross the English Channel.

A team of American specialists start to dismantle Me.262A-2a/U2 WNr 110555 - note the '555' just visible on the nosewheel door. Known as the 'Luftbomber' was the second 262 to carry the bomb-aimer in a glazed nose-section. (USAAF)

the front portion of each engine nacelle, so the Whizzers took several mechanics and flew to Lechfeld in a C-47 where they removed the needed components from another Me.262 . The damaged sections of the trainer were replaced and the aircraft loaded on board the aircraft carrier.

Administratively, there were other problems, as a memo from Lt Colonel A Detweiler to the Head of A-2 demonstrated. *'Airfield A-23 is currently being used by Exploitation Division, A-2 Section, this Headquarters, in connection with the project for shipment to Wright Field of approximately fifty flyable German aircraft. Due to the tecnical value of these aircraft, most of the experimental models and all of the most current German types and designs, it has been decided to deck load them aboard ship from A-23 in order that they may reach the United States with the least possible delay and in the best possible condition. It is essential that Airfield A-23 be retained under US control until these aircraft are shipped from that point.*

Since it is understood that Air Transport Command contemplates withdrawal from A-23 and that control of this airfield may be returned to the French on 1 July 1945, request that necessary action be taken by your section to insure complete access to this airfield during the period it remains under Air Transport Command control. This airfield is not released and returned to the French until such time as work in connection with the above mentioned project is completed. It is currently planned that shipment of these aircraft will have been effected from Airfield A-23 by 15 July 1945 at which time its use will no longer be required in connection with this project.

Watson's Whizzers were not the only group to make an epic journey across Europe according to the files of the United States Strategic Air Forces in Europe. On 3 June a request was made from the Office of the Commanding General to the Commanding General, Supreme Headquarters Allied

HMS *Reaper* in Cherburg Harbour awaiting the loading of the German aircraft.

Left: At least three Me.262s are on this barge along with a Fw.190 coming alongside HMS *Reaper* in Cherburg Harbour where (below) a Dornier Do.335A is seen being loaded. The jets were easier to handle as there were no propellers to be removed!

Expeditionary Force for clearance to fly four German helicopters from Stuttgart to A-42 Villacoublay. The four machines in question, a pair of Flettners and a pair of Focke Achgelis twin rotors were to be escorted by either a Piper Cub or C-64 Norseman

A pair of Flettner Fl 282 *Kolibri* 'Hummingbird' single-seat open cockpit intermeshing rotor helicopters assigned to Transportstaffel 40 at Mühldorf, Bavaria, had been captured by U.S. forces. One, recorded as V-32, was flown by test pilot Hans-Ehrenfried Fuisting, the other, recorded as V-12, was flown by Ernst-Willi Reiman. Focke-Achgelis Fa.223 *Drache* 'Dragon' V-14 was listed as being flown by Helmut Gersenhauer. Also on board was Frederick Will and Wilhelm Karl Otto Deilitz. The US intended to ferry captured aircraft back to the USA aboard a ship, but only had room for one of the captured Drachen. The RAF objected to plans to destroy the other, the V14, so Gerstenhauer, with two observers, flew it across the English Channel from Cherbourg to RAF Beaulieu via Lympne on 6 September 1945, the first crossing of the Channel by a helicopter. A second machine V-51, was flown by Otto Dumke, with Heinz Zelewsky acting as crew. Both machines were acquired by US forces during May 1945 at Ainring, Austria, where they had been in service with

Captured German aircraft sealed up for protection against salt corrosion aboard HMS *Reaper*.

Lufttransportstaffel40.

The documents record that there were concerns about flying short range machines across Europe, so three re-fuelling stops in France were arranged - Strassburg, Nancy and St Dizier.

Once all the aircraft had arrived at the port, Army Lieutenant Colonel 'Bud' Seashore supervised the loading sequence. He happened to be an old friend of Bob Strobell from his days at 1st Tactical Air Force headquarters.

During the *Reaper's* loading, each of the aircraft was given a protective coating to protect it from the sea spray. A shipping control number was assigned, and the planes were then placed on powered barges. Once they were in position, the aircraft were hoisted onto the deck of the carrier. This was repeated for nearly 40 aircraft of various types.

All the aircraft were finally inventoried, loaded and lashed to the deck of the H.M.S. *Reaper*, and the officers and men settled in for the long trip home. The carrier departed Cherbourg on 19 July 1945, bound for Newark.

There were forty-one aircraft aboard HMS *Reaper*, comprising ten Me.262s, five Fw.190Fs, four Fw.190Ds, one Focke Wulf Ta.152H, four Arado Ar.234Bs, three Heinkel He.219s, three Messerschmitt Bf.109s, two Dornier Do.335s, two Bücker B121 181s and the Doblhoff WNF.342 helicopter, two Flettner Fl.282 helicopters, one Junkers Ju.88G, one Ju.388, one Messerschmitt Bf.108 and one North American F-6. The latter was being shipped back to the USA under the instructions of Colonel George Goddard, to provide the manufacturer and Wright Field with an example of the latest state of 9th AAF field modifications to photo-recconaissance Mustangs.

Me.262A 777 *Jabo Bait*, seen at Newark on 24 August 1945.

Left: Me.262A by now marked as 666 Cookie VII is seen at Cherburg before shipping to the USA.

Below: the aftermath of the landing accident to the same aircraft at Pittsburgh Airport on 19 August 1945.

In all, about fifty aircraft were collected for transfer to the USA. Amongst the aircraft not taken to the USA was the Focke-Achgelis Fa 223E handed over to the RAF and the Heinkel He.111H flown to England and taken over by the 56th Fighter Group, 8th AAF. The balance of about ten aircraft may have included Heinkel He.162s and Messerschmitt Me 163Bs which later arrived in the USA aboard merchant ships.

Once at sea, there was little to do but relax and enjoy the voyage. Following the trans-Atlantic crossing, H.M.S. *Reaper* moored at Newark, New Jersey. The aircraft were then lifted by crane from the carrier deck onto barges. These were towed along a canal that bordered Newark Army Airfield where another large crane lifted each aircraft to the hardstand.

Once placed on the taxiway, the jets were towed to nearby hangars. They were then given local test flights prior to their ferry flights to Freeman Field or the USN Test Center at NAS Patuxent River. This work was done by many of the same people who had been involved with these aircraft in Germany and France, who had travelled with them aboard HMS *Reaper*, or had flown ahead of the ship on board the Ju 290 *Alles Kaputt* on its transatlantic ferry flight.

Roy Brown, Fred Hillis and Ken Dahlstrom were eligible for discharge upon their return, and all three left the project shortly after arrival in the States.

Most of the original team returned to civilian life and only Watson and Holt remained with the project long-term. Watson eventually returned to Wright Field while Holt was stationed at Freeman Field as the chief pilot for the 262 flight test programme.

Me.262A FE-110 is seen during a low-level pass at Freeman Field on 29 September 1945. *(NASM)*

As the US Navy had provided assistance in securing the British aircraft carrier to ship the machines across the Atlantic, five of the Me.262s were allocated to the Navy for their own evaluations at the Patuxent Naval Air Test Center and five remained in the hands of the AAF, and were ultimately flown to a remote airfield known as Freeman Field, Indiana, where the flight test programme could be conducted with relative secrecy.

Some of the aircraft, such as the short-range He.162s, or the Bf109s which were in too poor a condition to be made airworthy, were shipped to Wright or Freeman Fields by rail or road. The Me 163Bs, equally, could not make the journey under their own power, but many of the others were made airworthy and were flown, including most of the Me.262s. One of the Do.335s was test flown at Newark, but its rear engine overheated, and its pilot, Bell Chief Test Pilot Jack Woolams, was very lucky to get it back on the ground without the aircraft catching fire. The Do.335 then went by road to Wright Field.

On 19 August, two Me.262s were prepared for a ferry flight to Freeman Field via Pittsburgh. Ken Holt was to fly 666 *Cookie VII*, accompanied by Col. Watson in 444 *Pick II*. They departed Newark at around three in the afternoon, and planned to make a refueling stop at Greater Pittsburgh Airport en route to Freeman Field.

The flight arrived over Pittsburgh Airport about an hour later. The tower had previously been notified that the aircraft were not radio equipped and communicated with the jets using light-gun signals. The two aircraft circled

Arado Ar.234B *Snafu 1* seen at Melun before shipping to the USA aboard HMS *Reaper* an onward transportation to the US Navy at Patuxent River NAS. 'SNAFU' is an acronym for 'Situation Normal All Fucked Up' *(USAAF)*

the field twice, losing speed and altitude in the process.

Colonel Watson was in the lead aircraft, and he landed first at 1606 hours. After a short roll, smoke was observed coming from the front wheel. The control tower immediately notified the crash equipment to proceed to the Me.262. Watson turned right, off the active runway on to the grass and cut across to a nearby taxi strip.

Meanwhile, Holt turned onto his final approach which was observed to be fast, but otherwise normal. When he was approximately 20 feet over and one-third down the runway he was given a red light to go around. Holt was already committed to the landing, and the aircraft touched down just south of the main intersection at a high rate of speed. The tower then advised the crash equipment on the field to proceed after the second Me.262. Holt was observed to roll a short distance, then went off of the runway into grass. The jet continued on this track, paralleling the runway, until it dropped out of sight and burst into flames.

The impact tore off the landing gear and both engines, and broke the fuselage just behind the cockpit before the aircraft slid to a halt. The aircraft was a total loss. Salvageable components were fished from the wreckage, and the airframe was abandoned behind the airport fire station. Years later, it was covered and buried along with other aircraft wrecks.

On 28 September, Colonel Watson, and Bob Anspach accompanied test pilot Jack Woolams of Bell aircraft on a flight of three Me.262s from Newark (via Pittsburgh) to Freeman Field. They flew in loose V formation with Colonel Watson in the lead. During the flight, Watson's altimeter malfunctioned and instead of flying at 10,000 feet he had actually climbed to 16,000 feet without oxygen. Woolams knew that hypoxia would soon be setting in, and attempted to draw Watson's attention with hand signals. After landing at Pittsburgh, Woolams told Watson why he kept signaling for him to descend. The flight to Freeman Field then continued at 8,000 feet.

Holt was the only pilot at Freeman Field with orders authorising him to fly the captured jets. The T-2 Division at Wright Field would send Teletype orders to Freeman Field, calling for certain tests to be performed at certain altitudes, speeds, etc. Holt flew these missions, and forwarded the requested data via teletype back to Wright Field for analysis.

Holt had four Me.262s at his disposal, and he flew them all at one time or another. After an engine change or any major airframe maintenance, he would conduct the requisite test flights. Between the flight tests, maintenance checks and ferrying operations, he eventually logged over 200 flight hours in the aircraft.

Colonel Watson was involved in a near-fatal flight at Freeman Field in March 1946. Not long after take-off he discovered that many of his control inputs were having the opposite effect, and he quickly suspected that the elevator trim on the aircraft (FE-110) had been rigged in reverse. Through skillful handling he managed to get back onto the ground safely, thus averting a major disaster.

By May 1946, plans were in hand to shut down Freeman Field and transfer all USAAF, German, Italian and Japanese aircraft to storage facilities. Fighter aircraft were to be stored at the 803 Special Depot, Orchard Place Airport, Park Ridge, Illinois, where the newly-promoted Captain Strobell was charged with managing the inventory.

Meanwhile, the military formed a mobile ground display to make appearances at air shows and public events. The display included examples of the Me.262 (FE-110) as well as the Fw.190, V 1, V-2 and certain Japanese aircraft. Army Air Forces Base Unit 4140 (Research and Development Exhibition) was formed specifically to support these activities.

When not flying tests, Holt performed demonstration flights at several air shows across the country. He flew the two-seater Me.262 (FE-610) to an air pageant in Omaha, Nebraska on 14 July 1946. During a demonstration flight on the final day, one of the engines began losing turbine blades. He shut it down and landed without incident. The jet was then left at Omaha until the engine could be replaced, and Holt returned to Omaha and ferried the jet back to Freeman Field.

Once activities there began to be phased out, Holt also ferried Me.262s to Bolling Field near Washington D.C. and to Wright Field. The last remaining Me.262 at Wright Field was the A-1a/U3 FE-4012.

This aircraft was prepared for final tests against the Lockheed P-80 Shooting Star which involved removing the German jet's reconnaissance nose section, and replacing it with the fighter nose section from FE-111 which was aerodynamically cleaner. The aircraft was also given a high-gloss refinishing for the tests, and the designation changed to T-2-4012.

Following the tests, T-2-4012 was slated for a contract restoration with the Hughes Aircraft Corporation. It was delivered there in August 1947, and destined for Muroc Field upon completion. Once the restoration was finished, the Air Force apparently changed its plans for the machine, and Hughes was asked to store the jet. The company never made any attempts to fly it, and the aircraft eventually was turned over to an aeronautical school in Glendale as a static trainer.

General Hap Arnold ordered the preservation of one of every type of aircraft used by the enemy forces. In the end, Operation Lusty collectors had acquired 16,280 items that weighed some 6,200 tons, to be examined by intelligence personnel who selected 2,398 separate items for technical analysis. Forty-seven personnel were engaged in the identification, inspection and warehousing of captured foreign equipment.

ID	Variant	W/Nr	Remarks
			Watson's Whizzers aircraft details
000	262A-1a/U4	170083(V-083)	Trials A/C for 50mm cannon. Named *Feudin 54th A.D. Sq.* then *Wilma Jeannie*, then *Happy Hunter II*. Lechfeld - St Dizier - Melun 10 June 1945. Crashed en route Melun - Cherbourg.
111	262A-1a	unknown	Fighter. Named *Beverley Ann*, then *Screamin Meemie*. Allocated to Lt Robert Strobell. To USA aboard HMS *Reaper*. To US Navy as BuAer No. 121422. Preserved.
222	262A-1a/U3	unknown	Photo-Reconnaissance variant. *Marge*, then *Lady Jess IV*. Allocated to Capt. Kenneth Dahlstrom. To US Navy as BuAer No. 121443
333	262A-1a	unknown	Fighter. Named *Feudin 54th A.D. Sq.* then *Pauline*, then *Delovely*. Allocated to Lt Robert Anspach. Landed on Jersey 30 June 1945, then to Cherbourg, To USA aboard HMS *Reaper*. To US Navy as BuAer No. 121444.
444	262A-1a	unknown	Photo-Reconnaissance variant. Named *Connie the Sharp Article*, then *Pick II*. Allocated to Lt Roy W Brown. To USA aboard HMS *Reaper*. Ferried from Newark - Pittsburg - Freeman Field 19 Augst 1945 by Colonel Harold Watson. To FE-4012. Photo-Reconn nose exchanged for fighter version. Preserved.
555	262B-1a	110639	Two seat trainer. Named *Vera*, then *Willie*. Suffered nosewheel collapse at Cherbourg/Querqueville on 6 July 1945. Repaired. To USA aboard HMS *Reaper*. Preserved.
666	262A-1a/U3	500098	Photo-Reconnaissance variant. Named *Joanne*, then *Cookie VII*. Allocated to Capt Fred Hillis. To USA aboard

Testing in Ohio.

Frank Voltaggio Jr recalls one incident involving the testing of ME.262 Work Number 111711 at Wright Field, Ohio.

So far as is known this machine did not wear an 'FE-' or 'T2-' number but it is referred to in some documents as 'T2-711' but certainly wore '711' on the vertical fin. 111711 was surrendered at Frankfurt/Rhein-Main by defecting Messerschmitt test pilot Hans Fay on 31 March 1945. The Me.262 was on its first test flight from Hessental where it had been built by the Messerschmitt-controlled company, Autobedarf Schwabisch Hall. It was examined at Rhein-Main by USAAF Air Intelligence and then shipped onwards from Rouen, France to the USA aboard the merchant ship *Manawska Victory*.

After re-assembly at Wright Field, it was test flown by Russ Schleeh. After 12 flights there, totalling 10 hours 40 minutes. Its first two flights by

777	262A-1a	unknown	HMS *Reaper*. Crashed Pittsburgh Airport en route Newark - Freeman Field 19 August 1945 while being flown by Lt Ken Holt. Fighter. Named *Doris*, then *Jabo Bait*. Allocated to Lt William V Haynes. To USA aboard HMS *Reaper*. To FE-110.
888	262A-1a	500491	Fighter. Coded 'Yellow 7' of IV/JG.7. Named *Dennis*, then *Ginny H*. Allocated to Lt James K Holt. To USA aboard HMS *Reaper*. To FE-111. Preserved.
999	262B-1a/U1	11306	Two seat night fighter. Coded 'Red 6' of IV/NJG11. Surrendered to RAF at Schleswig and became 'USA 2'. Named *Ole Fruit Cake* after handover to Colonel Watson. To USA aboard HMS *Reaper*. To FE-610. Preserved at least until 1950s.
101	262B-1a	110165	Two seat trainer. Surrendered to RAF at at Schleswig and became 'USA 3'. Named *What Is It?* To USA aboard HMS *Reaper*. To US Navy as BuAer 121441. To Naval Research Laboratory, NAS Anacostia. Scrapped.
202	Ar234B	unknown	Named *Jane I*. To USA aboard HMS *Reaper*. To US Navy as BuAer 121445. Flown to NAS Patuxent River. Scrapped.
303	Ar234B	unknown.	Named *Snafu I*. To USA aboard HMS *Reaper*. To US Navy as BuAer 121446. Flown to NAS Patuxent River. Scrapped.
404	Ar234B	140311	Surrendered to RAF at Stavanger, Norway. Marked 'USA 40'. To USAAF as FE-1011 after arrival in USA.
505	Ar234B	140312	Surrendered to RAF at Stavanger, Norway. Marked 'USA 50'. To USAAF as FE-1010 after arrival in USA.

Russ Schleeh totalled 1 hour 45 minutes. Later flights were by Major Walter J. McAuley.

Frank Voltaggio Jr: *'I was with the Deputy for Intelligence beginning in January 1946, and in July of that year I was transferred (much against my will) to the satellite base operated by AMC at Clinton County Army Air Field near Wilmington, Ohio. This was known as the All-Weather Flying Center, and was involved in developing the technologies that have become somewhat common-place today for operating aircraft in bad weather.*

Housing of any kind was hard to come by for the military folks of that era, and was virtually non-existent in the vicinity of Wilmington. As a result, I chose not to give up my apartment in the northwest part of Dayton and commuted back and forth each day - a distance of about 40 miles in each direction. Mac and his wife, Liz, had an apartment in the same building, and we had become fast friends - especially the wives.

We knew that the German technology in jet aircraft was well ahead of our own during World War Two. You can include the technology of our war-time allies in that statement as well! As is well known and documented, the Germans had the Me.262 flying in combat in the latter stages of the war. They might have been able to turn the tide of the war had they been able to achieve this superiority earlier.

When the war ended there was a heavy effort to bring samples of the most advanced German military systems to the US, where the technologies could be studied in great detail. For aircraft and their sub-systems, the Air Materiel Command at Wright Field was the focus of this activity. The work was carried out by an organization known as the Deputy for Intelligence (T-2) that later became the Air Technical Intelligence Center and was the forerunner of today's Foreign Technology Division at Wright-Patterson Air Force Base.

The German jet aircraft carried a high priority for technical analysis, as did the V 1 and V-2 missiles. Two of the Me.262s were brought to Wright Field and put in the hands of the Flight Test Division headed by Colonel Al Boyd. He had a notable

Ju.290A-4 'A3+HB' *'Alles Kaput'* - formerly of KG200 - which flew from Paris Orly to Freeman Field via the Azores, Kindley Field Bermuda, and Patterson Field over the period 28 July / 1 August 1945. This aircraft was 'salvaged' -that is scrapped - in December 1946 at Wright Field. On dismantling a live demolition charge was discovered in one wing, but the detonator was defective and failed to fire the explosives!

FE610, a two seat night fighter variant of the Me.262.

crew of test pilots at the time, including some of the leading aces of the war.

One of the fighter test pilots in Col. Boyd's group was a neighbor and friend of mine by the name of Walt McAuley. Mac was one of the Air Force's first jet pilots, and he used to appear at air shows in one of the first F-80s, then known as the P-80. He spent a lot of time out at the base at the Muroc Dry Lake in California. Much of the flying by the Flight Test Division was done out there, including Colonel Boyd breaking the world speed record for aircraft while flying a P-80R.

One day in August 1946 I was making my usual run back to Dayton at the end of the work day. My 1936 Pontiac was chugging along Route 68, the main road between Wilmington and my first checkpoint at Xenia. I was cruising along about nine miles north of Wilmington when I passed a corner where the road from Spring Valley comes in from the northwest and meets Route 68. There was a combination gas station/general store on the northwest corner, and nothing but cornfields anywhere else within sight. When I passed the intersection I thought I heard someone call out my name - 'Hey Frank!' I slowed down and pulled over to the shoulder, got my car turned around, and went back to investigate. When I pulled into the station I immediately saw Walt McAuley. He was bare-headed and was wearing a light tan summer flying suit. There were blood stains on the suit in the area of the chest and shoulders and he was wearing a small bandage dressing at the point of his chin. My first thoughts were that Mac had been in an auto accident. But why the flying suit, which was never worn when on duty away from the flight line?

'Mac - what in hell are you doing out here and what happened to you?'

'Oh - I was up in one of the Me.262s and one of the engines caught fire and I couldn't put it out - so I bailed out.'

'Good grief - where's the airplane?'

'Oh - it's over in one of these cornfields.'

I asked him if he had been hurt at all and he said no - his 'chute worked fine and he wouldn't have had a scratch except that when he hit the ground, his chin hit the quick release knob for the parachute harness.

I offered to give him a ride back to the base but he said a staff car was already en route to pick him up. He asked me to tell his wife that I had seen him and assure her that he was still in one piece. His greatest concern appeared to be the confrontation with Col. Boyd that surely awaited him on return. He was not happy to be associated

Me.262A '711' seen on a test flight in the hands of Russ Schleeh over Ohio.(USAF)

711 at Wright Field.

with the loss of one of the colonel's pet aircraft.

When I got home to the apartment, Liz was with my wife and they excitedly began to tell me that Mac had been in an aircraft accident and they were awaiting further news on his situation.

Imagine their surprise when I told them - "Yes, I'm aware of it."

"How did you find out? Was it on the radio?"

"No, but I met him on the highway when I was driving home."

Needless to say, Liz felt greatly relieved when she learned that I had indeed seen Mac and could tell her that he was as hale and hearty as ever - albeit with a cut on the chin.

1st Lt Walter J McAuley, a test pilot for the Flight Test Division beside one of the AAF's first jet aircraft, the Lockheed P-80 'Shooting Star' in 1946.

She laughed at this, saying that it served him right - he was always avoiding the classes where they were giving instruction on how to land when you make a parachute jump.

The official account of the accident indicates that the Me.262 flight was being made in the company of an American P-80, so that a comparison of their respective speed capabilities could be made. After the speed run had been made at 20,000 feet altitude, the P-80 left to return to base and Mac stayed to make another run with the Me.262. After completing this speed run, Mac was about to throttle back when the aircraft suddenly began to vibrate violently. He observed grey smoke coming from the left engine, and on checking his instrument panel he noticed that the left engine tail pipe temperature guage was reading zero and was obviously not in working order. At about this time he also lost power in the right engine. He began a slow descent without power and contacted the base for a heading to return. While heading in the direction to the base, he attempted to restart the right

A Flettner helicopter is studied in France by the Americans. A number of German helicopters were evaluated. *(USAAF)*

engine. He got the engine started, but the flame would not extinguish, and the engine continued to torch and run at approximately 2500 rpm, with flame extending two to three feet behind the tail cone. The P-80 pilot, hearing of the problem on the radio, had returned to the scene and observed a long trail of smoke coming from the Me.262 as it descended.

When the altitude had dropped to 7,500 feet, Mac decided to abandon the aircraft and notified the P-80 pilot of his decision. He was seen to go out over the right side, and the plane was moving at 150 mph indicated, when he went out. The plane wasn't equipped with ejection seats like the modern fighters are. Mac's son, John told me that his dad told him he couldn't remember much about the bailout.

He told John he remembered opening the canopy and starting to put his hand out of the cockpit. The next thing he knew he was coming to with the parachute floating above him. After he hit the ground, he removed his helmet and saw that it had been heavily dented in the back. He theorized that he had been sucked out of the cockpit by the differential in pressure when the canopy opened, and had been hit in the back of the helmet by a part of the tail section, probably the horizontal stabilizer. The blow had knocked him out, or at least, had stunned him severely, and he didn't remember deploying the parachute.

When you think about it, Mac was lucky to get out of that plane alive. Crash helmets had only come into use by Air Force pilots with the entry into the jet age, and Walt just might have been the first American pilot to bail out of a jet aircraft.

Despite the crash, details of the tests of this aircraft were published in Technical Report F-TR-1133-ND, released in 1947. The report summarised Project No. NAD 29, *'Evaluation of the Me.262'*.

The report concluded that the '262 suffered from some poor features, including poor brakes. During the tests, '711 required five engine changes, which reflected the early stage of development of the jet engine, and the difficulties the German war machine experienced in obtaining suitable high-temperature materials. Although the handling characteristics of the '262 were considered poor, this was blamed on the fact that those flown had their aileron and elevator servo-tabs disconnected.

The overall conclusion was that 'T2-711' was superior to the average Lockheed P-80A in acceleration and speed, and comparable in climb performance, despite a weight penalty of 2,000 lbs. A maximum True Air Speed of 568 mph was measured at a pressure altitude of 20,200 feet.

Other machines tested

The Me262s were not the only captured German aircraft tested at Wright Field. One of the prototype Doblhoff helicopters WNF 342V4 - FE-4615/T2-4615 was surrendered by its designer at Zellam-See. It was shipped to Freeman Field, but was transferred to Wright Field before May 1946. Some tethered flights were made at Wright Field under the supervision of Dr Doblhoff, prior to August 1946. Further flight trials were considered, but these would have required engineering development to have been carried out on the helicopter.

Because developments akin to the Doblhoff design were already under consideration by the Kellett Aircraft Corporation for its own XR-17

The wreckage of the Me.262 buried in a cornfield south of Xania, Ohio on 20 August 1946. (USAF)

The fourth prototype Doblhoff WN 342 during a tethered flight at Wright Field in 1946 *(NMUSAF)*

helicopter, the WNF342 was sent to the General Electric Co at Schnectady, NY, for tests of its propulsion system in connection with the XR-17 project. In the meantime, Dr Doblhoff had visited General Electric and given it the benefit of his experience with the design concepts.

The XR-17 helicopter used similar rotor-tip propulsion principles to the Doblhoff design. This was later taken over by the Hughes Aircraft Company of Culver City, California. The WNF342 was used as a test-rig for component development to aid this development programme.

Other helicopters and rotary wing kites were also evaluated. A number of Focke-Achgelis Fa 330A rotor-kites were taken to Freeman Field and assigned to the Display Branch. Two were flown by Eastern Rotor Craft, which was contracted to carry out a full performance evaluation of the Fa 330 on behalf of Wright Field.

FE-4618/T2-4618 Focke-Achgelis Fa.330A was in store at Freeman Field by 19 June 1946 and was transferred to Wright Field during July. FE-4618 was test-flown at Wright Field, towed behind a truck. Four successful flights were made, followed by two in which the kite overturned on landing and was damaged. Previously, one of the Fa 330s had been rigidly mounted on a tiltable platform on a truck, which had been driven along a runway at speeds up to 35 knots to determine the adequacy of the rotor hub and blade strength.

During 1948, it was test flown from a USAF patrol boat in Tampa Bay, adjacent to MacDill AFB, Tampa, Florida as a means of carrying out extended

T2-4618 - a Fa.330 aboard a USAF Rescue launch in Florida. *(NMUSAF)*

flights of the Fa 330, which were not possible under tow behind a truck confined to the length of the Wright Field runway. Another factor was the possible use of the kite by the USAF as a means of extending the visual search range of small Rescue Boats.

The trials commenced in August 1948 and involved the construction of a platform on the aft deck of an 85-foot rescue launch. The kite was to be towed by a cable attached to a Navy Mark VII hydraulic winch installed on the boat. Several trials were conducted with the rotor kite tied down to the platform, before an attempt was made to carry out towed flight tests. The pilot was Capt Raymond A. Popson from the Flight Test Division at Wright Field. Unfortunately, the kite broke away as it was launched, due to a mechanical failure of the cable connection. After the pilot was rescued from the sea, the site of the crash was marked with a buoy, but this was stolen, or washed away in a hurricane, before the kite could be recovered. The trials were therefore terminated.

The test results were reported in Memorandum Report No. MCREXE-670-8-A of 31 December 1948, published by the Equipment Laboratory at Wright-Patterson AFB. Although the pilot reported favourably on the trials, they were not continued. At the time the project ended, two further Fa 330s were available to conduct further flights.

Horten aircraft
In the USA Jack Northrop had been building flying wings since the 1930s, and in the early post war years revealed that he was aware of the work of Walter and Reimar Horten who were German aircraft pilots and enthusiasts. Although they had little, if any, formal training in aeronautics or related fields, the Hortens designed some of the most advanced aircraft of the 1940s, including the world's first jet-powered flying wing, the Horten Ho.229.

The brothers wartime work was spent developing sophisticated and innovative tailless designs right through to their Ho.X. But it was the Ho.IX that was particularly impressive. This was a single-seat fighter-bomber of fifty foot span, powered by two BMW 003 jet engines, with a design maximum speed of 720 mph at 21,000ft, an endurance of four hours, fitted with a spring catapult ejector seat and armed with four 37mm cannon and 2,000kg bombs.

Walter (left) and Reimar Horten.

Above is the Horten Ho.IX VI glider. To the left is the unfinished Horteen Ho.220 V3 that was captured by American forces.

Built at the Aerodynamic Research Institute at Göttingen, the design was flight tested in glider form as the Ho IX VI. The second aircraft, the Ho IX V2, powered by a pair of Jumo 004s, was taken to Berlin-Oranienburg for flight testing, the first such flight taking place in January 1945. Unfortunately it crashed on 26 February during a forced landing following engine failure, killing its pilot. Having caught Hermann Goering's eye, the aircraft was authorised for manufacture by the Gotha company, its status being confirmed by its inclusion in the future German fighter building programme of 12 March, and was henceforth known as the Go 229.

Two months later the Gotha factory was captured by the American forces and the incomplete third aircraft, the Go.229 V3, was transported to the USA. Five further prototypes, including a fully-equipped production prototype, had been under construction at the war's end.

The V3 was larger than previous prototypes, the shape being modified in various areas, and it was meant to be a template for the pre-production series Ho.229 A-0 day fighters, of which 20 machines had been ordered. The V3 was meant to be powered by two Jumo 004C engines with 10% greater thrust each than the earlier Jumo 004B production engine used for the Me.262A and Ar 234B, and could carry two MK 108 30mm cannon in the wing roots. Work had also started on the two-seat Ho.229 V4 and Ho.229 V5 night-fighter prototypes, the Ho.229 V6 armament test prototype, and the Ho.229 V7 two-seat trainer.

A Horten glider and the Ho 229 V3, which was undergoing final assembly, were secured and sent to the United States for evaluation. En route, the Ho 229 spent a brief time at RAE Farnborough while it was considered if British jet engines could be fitted, but the mountings were

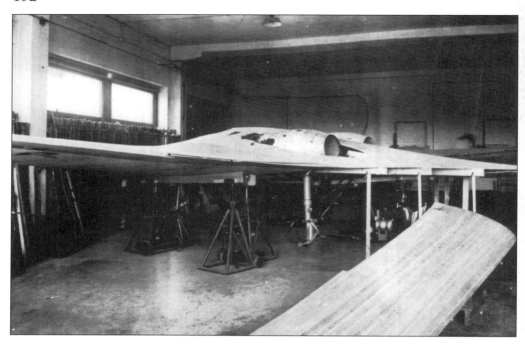

The Horten XIX V2 was powered by a pair of Jumo 0004-B turbojets, although it was originally planned to use a pair of BMW 003s, which forced a re-design of the engine bays to take into account the larger diameter of the more powerful Jumo units. The nosewheel came from a wrecked Heinkel He.177, including the retraction cylinder and hydraulics, the parts being provided by the Göttingen salvage organisation. Assembly took place in a three vehicle garage.

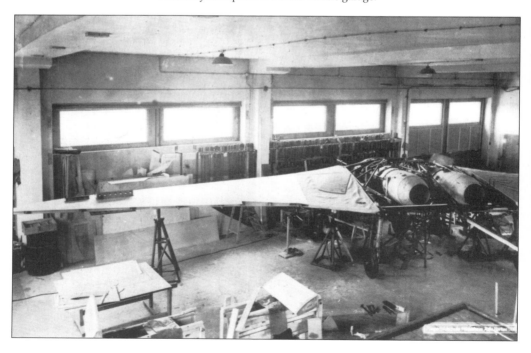

The first flight of the Ho.IX V2 was made from Oranienburg on 2 February 1945 with *Leutnant* Erwin Ziller at the controls - a flight that lasted for 30 minutes. Ziller made another flight next day, but deployed the aircraft's braking parachute too soon, which resulted in a hard landing that damaged the aircraft undercarriage. On the third flight, on 18 February, Ziller struck the ground while attempting to land, possibly with an engine problem. He was killed instantly when he was thrown out of the aircraft and struck a tree.

The incomplete and partially dismantled Horten flying wing arrives at Freeman Field by rail.

found to be incompatible with the early British turbojets only using centrifugal compressors with their comparatively larger diameter compressor sections, and not the slimmer axial-flow turbojet powerplants the Germans were using. The Americans were just starting to create their own axial-compressor turbojets before the war's end, such as the Westinghouse J30, with a thrust level only approaching the BMW 003's full output.

Recently documents have come to light from the Central Air Documents Office (CATO) of the Air Materiel Command at Wight-Patterson AFB in Ohio which reveal that the Americans were well aware of what the Horten brothers were doing, because they had an informant inside the company passing on information!

According to document EL-C 2763B T.I. 58618 in the CATO files, compiled by Lt Cmdr M A Biot and Lt Jayne, both of the USNR working with the US Naval Techical Mission in Europe: *'The information presented in this report was*

The incomplete Go.229 as it was photographed when captured by US forces following the overrunning of the Gotha Works at Friederichsroda on 14 April 1945 by the American 3rd Army's VII Corps. The Americans later assigned it the number T2-490. (USAAF).

Another Horten design that was evaluated was the Ho.VI research glider with an extremely high aspect ratio of 32.4:1. Constructed of wood and metal, the aircraft was considered by Walter Horten to be the highest performance sailplane of its day.

collected partly from captured documents and partly from an informant who was a draftsman for the Horten firm. Because of the technical limitations of the informant, some of the data must be accepted with appropriate reserve. The informant, however, had kept in continuous close touch with the designers, the Horten brothers, until the time of his capture. Most of the data obtained pertains to a new jet-propelled tailless fighter, the H-IX-V2, which was designed and built at Gottingen under the technical supervision of Major Walter Horten and Oberleutnant Riemar Horten of the Luftwaffe. In the following sections are given a short description of the H-VIII, a large commercial flying wing which is under construction at Gottingen, and a description of other Horten designs. The last section is a complete list of captured documents which are being processed and filed.'

So, it seems that the Americans had a spy inside the Horten organisation prior to wars end. More recent reports from uncredited origins seem to suggest that the H.VIII was to be in two forms; the A model was a long, smooth blended wing, its six jet engines buried deep in the wing and the exhausts centred on the trailing end. Resembling the Horten Ho.229 flying wing fighter, there were many odd features that distinguished this aircraft; the jettisonable landing gear and the wing made of wood and carbon based glue, are but two. The aircraft was first proposed for the Amerika Bomber project and was personally reviewed by Hermann Goering, but after review, the Horten brothers - with deep dissatisfaction - were forced to share design and construction of the aircraft with Junkers and Messerschmitt engineers, who wanted to add a single rudder fin as well as suggesting underwing pods to house the engines and landing gear.

The B model was generally the same as the A model, except the four

The Darmstadt-Munchen DM.1 glider seen at Munich Prien airfield in May 1945.

Two more views of the DM.1, showing clearly the extraordinary thickness of the delta aerofoils. *(USAAF)*

(down from six) engines and four-wheel retractable landing gear were now housed in underwing pods, and the three-man crew housed under a bubble canopy. The aircraft was to be built in huge concrete hangars and operate off long runways with construction due to start in autumn 1945, but the end of the war came with no progress made. Armament was considered unnecessary due to the expected high performance.

However, neither of these descriptions match with what was said in the US Navy Report: 'This airplane is under construction at the Luftwaffe Sonderkommando 9 at Gottingen. It should be ready for flight testing around November, 1945. It has a span of 157 feet, and is powered by six 600 hp BMW pusher engines.

The airplane range is computed to be 4500 miles at a cruising speed of 200 mph and at an altitude of 5,000 feet. It will carry about 60 passengers. The center section is of welded steel tubes. The outer wings are of wood with one main spar and one auxiliary spar. The control system is the same as for other Horten designs. Powered controls are envisaged. The nose wheel is not steerable but self-centering by spring

The DM.1 was later modified with the canopy from a P.80 and tested in the NACA's wind tunnels.

and cam. There is no pressure cabin, The adaptation of a venturi of three meters diameter under the wing for use as a flying wind tunnel has been proposed.'

Once the aircraft was safely in storage at Freeman Field and allocated the FE-490/T2-490 serial, a close study revealed that it would take an estimated 15,000 man hours to make it flyable - but this was not done.

One of the more exotic aircraft tested by the Americans was the Darmstadt-Munchen DM.1. The performance of the Me-163 encouraged Lippisch to experiment with supersonic flight and he created several designs that culminated in the DM.1 glider. Lippisch intended to test the DM 1 to determine the handling characteristics of a sharply-swept delta wing aircraft flying at low speeds, and then he planned to add power and push the aircraft to higher speeds. He optimistically hoped to reach Mach 6. This was intended as the aerodynamic test vehicle for the Lippisch P 13 ramjet-powered fighter. Construction began in August 1944 at the Flugtechnische Fachgruppe (FFG) Darmstadt but the war ended before workers could finish the glider and the Allied armies discovered it when they occupied the base at Prien am Chiemsee in southern Germany early in May 1945. After the airframe had been completed under American supervision, it was crated for shipment to the US under the orders of Theodore von Kármán.

The DM1 was taken to the National Advisory Council for Aeronautics (NACA) facility at Langley Field for full-scale wind tunnel testing. It was crated and sailed from Rotterdam on the SS *King Hathaway* bound for Boston, where it arrived on 19 January 1946. The authority for the wind tunnel tests was dated 5 February 1946. The tests were completed by January 1948 and were of use in the development of the Convair XF-92, which was the prototype from which the F-102 was developed. The wind tunnel tests were quite extensive and included tests in the aircraft's original configuration, with its large fin removed, and then in both of these conditions with modified wing leading edges designed and installed at Langley.

It had been intended to mount the DM.1 on top of a Douglas C-47 carrier aircraft which would have taken the aircraft to a height from which it could have been released for piloted free-flight. However, the wind tunnel tests showed that the aircraft was directionally unstable and had higher drag and a significantly lower lift/drag ratio than predicted. It was concluded that aircraft of the delta configuration would always have to be landed under power. It was considered that free-flight tests of the glider would have been unsafe and therefore no such trials were carried out.

Details of the wind tunnel investigations were given in two NACA

Two views of the incomplete Me.P1101 before shipping to the USA. (*both USAAF*)

reports, NACA RM No. L6K20 of 12 February 1947 and NACA RM No. L7F16 of 5 August 1947, the first reporting on the 'Maximum lift and flow characteristics of the triangular wing planform' and the second on 'Maximum lift and stability characteristics'.

Within nine days of the 15 July 1944 issue of the design specifications for the Emergency Fighter, the Messerschmitt design bureau under Dr. Woldemar Voigt had formed a preliminary paper design for the P.1101. The aircraft which was developed initially had a short and wide fuselage, tricycle landing gear, mid-mounted wings with an inner sweep of 40° near the fuselage, and a shallower 26° outboard. The single HeS 011 jet engine was to be mounted within the fuselage, being aspirated by two rounded intakes located on either side of the cockpit. The high tail was of a V configuration, and mounted on a tapered boom which extended over and past the jet exhaust, while the cockpit was forward mounted, with the canopy integrated into the fuselage and forming part of the rounded nose of the aircraft.

By late August 1944, the design, still in paper form, had evolved into a sleeker incarnation, with the previously stout fuselage lengthened and narrowed with a conical nose section added in front of the cockpit. The compound sweep wing was also abandoned, the outer wing of the Me.262 instead being adapted. Proposals for a pulsejet and rocket combination, the P.1101L, were also put forth. The design was further developed, including a longer nose, and after the wind tunnel testing of a number of wing and fuselage profiles, the decision was made to undertake the construction of a full-scale test aircraft. This finalised design and associated test data were submitted to the Construction Bureau on 10 November 1944 and the selection of production materials was begun on 4 December 1944.

On 28 February 1945, the RLM settled on a competing design, the Focke-Wulf Ta.183, as the winner of the Emergency Fighter program. This decision was based in part on the considerable design difficulties being encountered by the Messerschmitt P.1101 design team. For example, the cannon installation was proving too crowded, the mainwheel retraction and door mechanisms were too complex, the fuselage needed a great number of 'strong points' to deal with loads, and the anticipated performance had fallen below the RLM specifications due to increased weight.

Since considerable work had already been done on the P.1101 design, the RLM decided to continue reduced funding for Messerschmitt to carry out experimental flights testing the swept back wing at anticipated speeds up to Mach 1. The worsening war situation led to the expedited, but risky, approach of building a full-scale prototype in parallel with detail construction and continuing statistical calculation, while existing components such as the wings (Me.262), landing gear (extended Bf.109), and flight components were utilized where feasible. It was also intended for the test flights to be conducted with 35, 40, and 45 degree wing sweep. Production of the V1 prototype was begun at Messerschmitt's Bavarian Oberammergau Complex with a projected first flight in June 1945.

The P.1101 V1 prototype was of duralumin fuselage construction, retained the outer wing section of the Me.262 but with larger slats and, as mentioned previously, the wing sweep could be adjusted on the ground from 30, 40, to 45 degrees; this was for testing only and never intended as an operational feature. The fuselage-mounted tandem intakes of the preliminary designs were replaced by a single nose intake, and the canopy became a bubble design, which afforded better all-around vision than the initial integrated canopy. The production prototype also incorporated a more conventional swept tail design, which was constructed of wood and remained mounted on the tapered tail boom. A T-tail was also designed. The tricycle undercarriage consisted of a steerable, rear retracting nose wheel and long forward-retracting wing root mounted main gear. The prototype was fitted with an apparently inoperable Heinkel He S 011 jet engine, but given the non-availability of this engine, a Jumo 004B was fitted for test

The P1101 re-assembled in the USA and awaiting use as a pattern from the Bell X-5. (USAAF)

flights. (Changing the type of engine was meant to be comparatively easy.) In addition, the production model was to be equipped with a pressurised cockpit and armoured canopy, and to be armed with two or four 30 mm MK 108 cannons, Ruhrstahl X-4 air-to-air missiles, or both.

By the time an American infantry unit discovered the Oberammergau complex on 29 April 1945, the V1 prototype was approximately 80% complete. Wings were not yet attached and appear to have never had skinning applied to their undersides. The airframe was removed from the nearby tunnel in which it was hidden and all associated documents were seized. There was some lobbying by Messerschmitt Chief Designer Woldemar Voigt and Robert J. Woods of Bell Aircraft to have the P.1101 V1 completed by June 1945, but this was precluded by the destruction of some critical documents and the refusal of the French to release the remaining majority of the design documents (microfilmed and buried by the Germans), which they had obtained prior to the arrival of American units to the area. The airframe meanwhile became a favourite prop for GI souvenir photos. Later the prototype was shipped first to Wright Patterson AFB, then to the Bell Aircraft Works in Buffalo, New York in 1948. Damage ruled out any possibility for repair although some of the Me P.1101's design features were subsequently used by Bell Aircraft Corporation of Buffalo, NY, for study during the development of the X-5.

An initial proposal made by Bell was to replace the Heinkel-Hirth He S-011 jet engine with an Allison I-35, to form the basis of a new interceptor design. The studies also included a proposal to replace the wing centre-section of the P 1101 with a new one which would permit in-flight changes of the sweep angle. The P 1101 had a wing which could be changed from 35 to 45 degrees on the ground, with three fixed positions. Bell also noted that the simple layout of the P 1101 permitted changes of engine installation with the minimum of effect on the fuselage structure. Bell proposals suggested the use of the P 1101 as an engine test-bed. Outline proposals were made for the installation of J34, J35, J46 and J47 turbo-jets, a ramjet, and a liquid fuelled rocket engine as alternatives. Further studies, allied to the damage which the P 1101 had received during ground handling, led to the abandonment of these proposals. The proposal to fly the P1101 was abandoned and the aircraft is believed to have been scrapped, because it was more expedient to build a new aircraft embodying the P 1101 design principles. The Bell X-5 was an enlarged development of the original concept, with provision to vary the wing sweep in the air - in the P1101, the wing sweep could only be adjusted on the ground. The company formally proposed the construction of a variable-sweep research aircraft on l February 1949 and on 26 July 1949 a contract (W33-038- ac3298) was placed for the supply of two Bell X-5 aircraft. These had the Air Force serial numbers 50-1838 and 50-1839. The first X-5 made its first test flight from Muroc on 20 June 1951, with the second flying on 10 December 1951.

Chapter Eight
BIOS, CIOS, FIAT And Other Acronyms

At the same time as aircraft and equipment were being recovered, other organisations were creating and publishing a myriad of documents on their investigations and interrogations.

The three main - but not only - organisations were the British Intelligence Objective Sub-Committee - BIOS, the Combined Intelligence Objectives Sub-Committee - CIOS and the Field Information Agency Technical (United States Group Control for Germany) known as FIAT which was a US Army organisation. Clearly BIOS documents originated from the British, whereas FIAT and CIOS reports were both based on joint British and American investigations, covering German Science and Industrial Institutions.

These reports on German engineering industry were published between 1946 and 1949. BIOS were operating under the auspices of the British Foreign Office – is known to have published over four thousand reports which are lodged in un-digitised form in The National Archives at Kew.

This collection includes reports on technology, engineering, and industries, mainly in Germany, during and immediately following the Second World War. The information was collected by various governmental intelligence agencies; and reports cover subjects such as the German clock and watch industry, chlorine plants, pharmaceuticals, viscose rayon plants, radar, and chemical industries. They were by no means restricted to aviation topics – but those which were, were reported on by personnel from the RAE and MAP.

Topics ranged from *'The Vioth-Schneider Propeller'* (BIOS Final Report 28) - a maritime propeller that gave variable direction thrust as well as pitch; the self explanatory *'Some Aspects of German Work on High Temperature Materials'* (BIOS Final Report No. 272) or CIOS File No. XXX-15 *'The AGFA Film Factory, Wolfen'*.

Another typical report was the *'German Aircraft Industry, Dornier Werke, Friedrichaven area'* (BIOS Final Report 30) issued following a four man team - three from the MAP and one from RAE who visited from 17-28 July 1945 - *'...The object of the visit was to investigate the productive capacity and production methods of the Dornier Company who were engaged on the design and production of aircraft'.*

The Allies were particularly interested in just how German industry had made such advances in technology when defeat loomed ever closer and shortages grew exponentially.

An example of this was a report issued by the Combined Intelligence Objectives Sub-Committee that contained an interesting review of the production methods used at various dispersed Messerschmitt works. Among the factories visited were:

Oberammergau: centre of engineering and experimental activities.
Fischen : centre of master-tool manufacture.

Oberstdorf: the master-tool repository.
Eschenlohe: location of wind tunnel. Fuselage components.
Kottern: the main tool-manufacturing plant.
Laubas: the pattern-making shop.
Fischenmuhle: machine cutting-tool manufacturing plant.
Kematen: flap and aileron-assembly plant.
Schwaz: jet-unit cowl-assembly plant.
Horgau Forest: wing-assembly plant.

Much of the German manufacturing facilities had been transferred to underground caverns and tunnels to escape from Allied bombing. *(USAAF)*

The methods discovered did not show any outstanding differences from that of the Allies, athough as might be expected the shortage of certain materials made necessary the development of substitute or makeshift processes. During the course of the interrogation Herr A. Stempfie, the tooling chief at Langenwang, Fischen and Oberstdorf, described the Messerschmitt practice in connection with the various forms of press tools.

Blanking dies were made of wrought iron plate. The working edges of the dies were bevelled and Arkaton metal deposited by arc welding to give the necessary cutting edge. Arkaton was a patented product made under licence from the Daimler-Benz Company and after application was very nearly glass hard with a Rockwell C reading of about 55 to 60. Tool steel was used only for very small blanking dies.

Owing to the shortage of cobalt and wolfram, tool steel containing these elements was not used during 1944 and 1945. For short runs a low carbon steel, flame hardened, was used.

Blanking dies were repaired by removing sections of Arkaton and re-welding and were made in sizes up to ten feet in length by three feet wide.

The practice of using mild steel punches with zinc dies was not adopted owing to the scarcity of zinc, but the blanking dies had, it was stated, a long life and quantities of from 10,000 to 20,000 parts were produced with them. The maximum thicknesses of material blanked out by these tools were about 11 to 8 SWG (Standard Wire Gauge) in light alloy and about 14 SWG in steel.

Rubber-press forming was used for about half the parts made. Pressures of from eight and a half to twenty-one tons per square inch were used. The necessary allowance for spring-back was established by trial and error. For long runs, cast-iron forming dies were used and compressed impregnated wood for small quantities. Light alloy was heat treated before the forming process and was torch heated at points where the forming operation was particularly severe.

For deep drawing the usual type of double-action tool with pressure-pad was used, but in special cases an application of extrusion forming was adopted. In such cases a tool with a highly polished punch and die was made, the space between the two being carefully calculated. An aluminium slug was placed in the bottom of the die and under a punch pressure from twelve and three quarters to seventeen tons per square inch. The material was squeezed up and round the punch.

A mechanical stripper removed the part from the die. This method of forming was only used where the bottom area of the part to be formed did not exceed three to four square inches. Compound dies to perform four to five operations were used very extensively. Plastic-punch press tools were not used. An almost complete absence of zinc and lead prevented the use of the drop-hammer process of forming.

In the manufacture of assembly jigs, pipe and structural-steel sections were all used, but the use of pipe predominated. There were eight standardised sizes of pipe - 24, 56, 84, 96, 116, 180, 214 and 360 mm in diameter. For each size of tube there were standardised sets of castings for building them up into structures, the system being rather similar to those used in various constructional toys. The use of welding was by this method practically eliminated.

Only rough setting of the location blocks on the fixtures was done at the tool manufacturing plant and the final setting was carried out at the factory where the components were to be actually produced. Final settings were made by transit or by height gauge and precision measuring tools. The frame of the fixture was not normalised before the final setting of the locations, as it was not considered to be necessary in view of the fact that welding was not used in the construction of this type of fixture.

Master reference gauges were frequently used in final setting operations. When assembly fixtures were scrapped owing to a change of design or introduction of a new type, about 70 per cent of the material used in their construction was salvaged for re-use. Formers and location blocks on assembly fixtures were often made of plastic materials or improved wood.

Pre-production methods did not reveal any particularly advanced technique. Approved design drawings and layouts were sent to the template division, where male and female templates were cut out of sheet steel approximately 0.094 inches in thickness. A No. 1 master template (female) was made, known as the negative. This template was painted yellow and remained at all times at the Oberstdorf tool repository.

A No. 2 master (male) template was produced by the use of the No. 1 and was called the positive. It was painted orange and was used in making the assembly fixtures and acceptance gauges. The No. 3 master (female) was used as an acceptance gauge, painted red, and produced by the use of the No. 2 master.

Master gauges were made from tubing welded and machined. In large master gauges the standardised tooling materials were used to a large extent, often in combination with the welded tubing. Tolerances maintained on No.

Original caption: 'Entrance to the V-1 and V-2 underground factory near Nordhausen, Germany, where hundreds of jet and rocket assemblies were found by US Ground Forces. A US Army 9th Air Force cameraman accompanying advance US ground elements, made this photo'.

The Dora/Mittelwerk site had much that could be exploited by the Allies. (USAAF)

1 master were +/- 0.004 to 0.008 inches in a length of 23 feet. For No. 2 and 3 masters a limit of +/- 0.012 inches was maintained. Final setting of location points was effected by the usual practice of bolting and dowelling.

At the Oberammergau factory two interesting developments were noted. One was an experimental riveting machine which first drilled a hole in the sheets, inserted and cut the wire rivet-stock to proper length and finally headed it on both sides of the sheet with a good flush-type head. This machine was capable of finishing 20 rivets per minute but had given much trouble and only two had been built.

Another unique and accurate piece of equipment that had been developed was for profiling wood patterns directly to compound curvatures from design loft lines and generated much interest.

In arrangement, the machine was of fairly conventional planer type with a reciprocating table to carry the work-piece, and a cross-rail slide on which the tool could be traversed in each direction. Vertical adjustment of the tool could also be made.

The main interest of the machine and the manner in which it differed from other types designed for the same general purpose was that it was possible to copy a profile directly from a mould-loft or other template layout without the need for first making a three-dimensional model to serve as a master. The machine was primarily intended for the rapid manufacture of master patterns and form blocks in wood, improved wood and light alloys, especially magnesium-based materials such as Elektron.

For this purpose a layout drawn accurately to scale and representing regularly spaced mould lines was attached vertically to the side of the machine table. It moved, therefore, with the same speed and direction and in the same plane as the work surface.

An optical system arranged on one of the side columns of the machine scanned the layout and projected an image of the line, enlarged 7.5 times, on to a ground-glass screen in front of the operator, who was seated on a raised platform facing the column and worktable of the machine. To the operator, the lines appeared to move across the screen as the workpiece on the table was fed under the cutter. A follower with the same proportions as the cutter was placed in front of the first lens of the optical system and its image was superimposed on that of the lines as they were reflected on to the screen.

By suitable manipulation of the appropriate controls, the operator kept the image of the follower in contact with that of the line. Hydraulic linkage simultaneously raised and lowered the cutter head and front lens of the system and as the lines are followed on the screen they were automatically reproduced on the surface of the work-piece. As a magnification of 7% was used, an error on the part of the operator in following the lines on the screen was reduced, on the work, to an amount less than one seventh of the original divergence.

A spherically ended cutter was used, driven at speeds ranging from 6,000 to 12,000 r.p.m and the cutter speed was 800 mm per minute. The transverse position of the cutter was also controlled hydraulically and, through a selector control, it could be moved by increments equivalent to those required by the layout.

The complete contour of the block was established by machining a series of grooves to the appropriate depth, each groove corresponding in curvature with a particular layout contour line. When the longitudinal cuts had been completed the block was turned horizontally through 90 deg and a second set of contours was machined in conjunction with another layout on which transverse sections were represented.

When machining had been completed, the block was unclamped from the machine and the high points left between successive cuts removed down to the full depth of the grooves and blended to give a smooth surface of

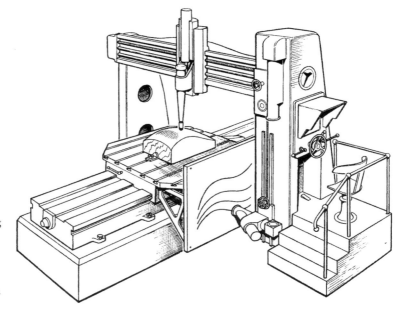

A sketch of the Messerschmitt three-dimensional profiling machine that so fascinated the Allies. It was built on the principal of a plano-mill and only machined in two axis at any one time.

It was realised by the Allies that sooner than exploit the hardware already produced, it would be better, and in the long term more financially profitable to exploit the German Patents for all this high technology. *(USAAF)*

compound curvature.

A comparatively simple optical system was used, consisting of only two lenses, four plane mirrors and a ground-glass screen. Illumination of the layout was effected by a lamp contained in a housing which was an integral extension of the box enclosing the front object lens and the first diagonal plane mirror. Light reflected from the surface of the layout was collimated by the front lens from which it emerged as a parallel beam and was reflected vertically upwards by the first plane mirror. It then passed to the second object lens which focused the image via the second, third and fourth plane mirrors on to the screen.

Collimation of the light by the first object lens was an ingenious feature in the optical design, as it overcame the difficulty of maintaining focus and magnification in spite of the variation in the total length of the optical axis. This variation was unavoidable as the front lens had to move vertically in order to follow the rise and fall of the layout contours, but the viewing screen itself remained fixed. The effect, therefore, in an optical sense was that the layout surface was placed at infinity and remained so, irrespective of the separation of the first and second object lenses. It was always projected in focus from this (imaginary) distance on to the screen.

Other teams, such as a group from the British Non-Ferrous Metals Research Association studied the process of pressure welding light-alloy sheets without fusion, in which a degree of success had been recorded. This came about through discovery by the RAE and MAP of certain Junkers aftercoolers during the war. Metallurgical examination of the German method of pressure welding revealed that this had been developed at the Junkers Flugzeug-und Motarenwerken A.G. by a pressure-welding process. It was confirmed when MAP investigators were able to visit German factories and technical notes on the method used came to light.

Two way traffic

The flow of technicians, equipment and scientists was not just one way from Germany, and it was not just governmental and the military. One of the members of Von Kármán's Scientific Advisory Group was Boeing's chief aerodynamicist, George Schairer. During his visit to Germany, Schairer examined data obtained by German aircraft manufacturers on the advantages of swept wings, and became so convinced of the merits of such a design that in May 1945 he wrote a letter to Boeing management suggesting the matter be investigated. The NACA wind tunnel tests showed that the model under consideration for the Air Force suffered from excessive drag. Boeing engineers then tried a revised design, the Model 432, that had the four engines buried in the forward fuselage, but though the Model 432 had some structural advantages, changing the engine layout didn't really reduce drag very much. The Boeing engineers turned to the swept-wing data obtained from the Germans and promoted by Schairer. A little design work by Boeing aerodynamicist Vic Ganzer led to an optimum sweepback of 35 degrees.

Boeing then modified the Model 432 design with swept wings and tail, resulting in the Model 448, which was presented to the USAAF in September 1947. The Model 448 had the four TG-180s in the forward fuselage as had the Model 432, plus two TG-180s buried in the rear fuselage. The Boeing project manager, George Martin, had decided that the company's entry into the bomber competition needed greater range and performance, and that led to six engines rather than four.

After many more refinements and re-works, what emerged was the Boeing B-47 jet bomber, which was a major postwar innovation in combat jet design, and led to the development of modern jetliners. The B-47 also made use of another German invention, the ribbon braking parachute. While it never saw serious combat use, the B-47 was the mainstay of US strategic defence in the 1950s, over 2,000 being built.

Another example of how German technology was used - and in this

German equipment was abandoned everywhere - the Mittelwerk tunnels were almost clogged. *(USAAF)*

Original Caption: *German Heinkel He 219 night fighter equipped with radar antennae on nose. Ship was captured intact at Bindbach, near Bayreuth. On field taken by 11th Armored Division of US 3Rd Army. (USAAF)*

instance in a much more direct way - was that of de Havillands and Rolls-Royce.

Munich and its surroundings were heavily bombed during the war. The Bayerische Motoren Werke at Oberweisenfeld in the locality suffered considerable damage during these operations, but when this area was overrun by the Allied armies in 1945 it was found that the special high-altitude test plant for aircraft engines situated on the northern boundary of the BMW factory was still in use.

Building of this plant had been commenced in May 1940, and heavy bomb damage had been sustained in May 1944. It had, however, been repaired by October of that year and from then until the collapse of Germany in 1945 the plant had been in regular service.

This plant was in the American occupation zone, but arrangements were made by the Ministry of Aircraft Production for its use by British manufacturers, and both the de Havilland Engine Co Ltd., and Rolls-Royce Ltd., were given facilities for testing engines there.

The de Havilland Company were approached by the Ministry of Aircraft Production in July 1945, and after a preliminary survey of the plant by one of the company's engineers, arrangements were quickly made to test the Goblin I jet unit in the plant.

At that time no other facilities existed for testing engines under conditions comparable with those obtainable at Oberweisenfeld. The test cell had a diameter of twelve feet and could be evacuated to pressures simulating conditions at a height of 50,000 feet. Air could be supplied to the intakes at a speed of 550mph and temperatures as low as -65 deg C could be maintained.

A special mounting for the Goblin engine was necessary but all the essential equipment and arrangements for the test were completed within twenty-one days. Another period of fourteen days was required for installation at the plant. A series of tests was then made at different speeds and under conditions corresponding to altitudes ranging from ground level to 43,000ft.

The tests were completely free from trouble of any kind, though, owing to the very heavy power consumption of the plant, it was possible to operate it only at night.

The complete series of tests on the Goblin I required a total of 42 hours running and the engine was not examined until the tests had been completed. After being returned to England, the unit was re-calibrated and the output was found to be substantially the same as before. When it was stripped, however, inspection showed that a great deal of dust had been drawn in from the heavily bombed area which surrounded the test plant and a certain amount of superficial damage had been caused.

The success of the tests and the value of the information obtained led to the decision on the part of the de Havilland Company to undertake a second series, this time on a Goblin II engine. These tests included observations of

The D.H. Goblin jet engine in the test cell of the BMW high-altitude test plant at Oberweisenfeld

combustion and flame stability at altitude, and engineers from Joseph Lucas, Ltd., took part.

Normally, the method of observing the engine during test was through a periscope in the side walls of the cell, but for the purpose of these special combustion tests quartz windows were inserted in a flame tube of the engine and windows were cut in one of the explosion doors at the top of the test cell. Gas samples were also taken from the exhaust by the Lucas engineers. During the tests the inside of the cell was brilliantly illuminated and formations of fog and ice under high-altitude conditions could be readily observed. In taking measurements of the engine thrust it was found necessary to supply glycol to the sliding joints to prevent them from freezing.

The plant itself ran extremely well and was remarkable for flexibility, steadiness and stability under all operating conditions. The Siemens automatic control system, housed in a separate room, was of outstanding interest. In spite of its complexity it was surprisingly reliable. Constant altitude conditions were maintained irrespective of the air consumption and the test engineer was therefore relieved of all anxiety on this score.

Consideration of German Patents.
Given the plethora of investigations and report publishing by the British, it is not surprising that there was much speculation as to how German patents should be dealt with. The British Engineers Association issued a memorandum which made a number of well-reasoned recommendations.

In framing these recommendations it was considered that they should not tend to discourage or prevent scientific or technical developments in this country, should be calculated to benefit British industry to the greatest possible extent, should free from German control processes essential to the well-being of this country and should contribute to the extraction of

So much equipment, just abandoned and wrecked. *(USAAF)*

reparations. It was assumed in any case that German interests in United Kingdom patents would be sequestrated.

There were two possible alternative policies for dealing with sequestrated patents. They could be made available to British industry by licensing, sale or otherwise, or they could be cancelled, leaving the inventions to which they related at the free disposal of the world.

It was recommended by the British Engineers Association that the latter course should be adopted as a guiding principle, except where cancellation would injure some non-enemy interest.

It appears, however, that it was necessary for the Allies to take corresponding action and, indeed, that neutral countries should agree to do so: otherwise the situation might arise that a German patent cancelled in this country was valid in some other country, so that an infringement of patent rights would follow in the case of a British article being exported to the other country. Until international agreement could be reached German interests should be sequestrated.

The report states:

On the basis of these principles the memorandum recommends :-

(1) That German-owned patents in this country, which lapsed during the war should not be reinstated.

(2) Those German patents which have been kept in force should be cancelled where there is no non-enemy interest.

(3) Where a patent has been kept alive by a non-enemy licensee it should be sequestrated but preserved for the protection of the licensee.

(4) Any German interest in patents which are part-owned by Germany should be sequestrated.

(5) Patent applications of German nationals at present on the files in this country should be made available to British industry, whether they have been used during the war or not, unless publication would be inadvisable on grounds of national security.

(6) All information regarding inventions made in Germany during the war should be made available to British industry. The Government have already announced that inventions made in Germany since 1938 shall not form the basis of patents in the United Kingdom, and this has been embodied in a Bill to amend the Patents and Designs Act, 1927.

(7) Germany should be compelled to declare and render-up patents taken from allied owners and also royalties due for working the patents during the war.

(8) British subjects should be enabled to reinstate patents in Germany which lapsed during the war.

(9) Germany should be made to assign to the allies all German-owned patents in neutral countries, or, alternatively, arrangements should be made with these countries to cancel patents covering inventions which have been freed from protection elsewhere.

From this came calls for an international conference to consider the question of the future treatment of German-owned patents in Allied countries.

Delegates attended the London meeting from Australia, Belgium, Canada, Czechoslovakia, Denmark, France, Luxembourg, Netherlands, Norway, Union of South Africa, United Kingdom, and the United States of America.

Patents taken out by Germans existed in varying numbers in all countries of the world. Complete unanimity prevailed among the Allied nations that in no circumstances should any such patents within their territories revert to the former German owners, and the question as to how such rights should in future be disposed of presented many difficulties. A strong sentiment prevailed that it would be unfortunate if the continued existence of these patents would constitute an obstacle to international trade.

As a result of the discussions at the conference, the representatives of France, the Netherlands, the United Kingdom and the United States of America signed an accord. This had the effect of making all patents of former German ownership then controlled by their Governments, and in which there was no non-German ownership, and in which there is no non-German interest existing on 1 August 1946, available within their respective territories to all nationals of those countries without payment of royalties or without any requirement to manufacture within the country where the patents exist. The representatives of Australia, Canada, Czechoslovakia, and the Union of South Africa agreed to recommend to their respective Governments that the accord should also be signed on behalf of those Governments.

Representatives of Belgium, Denmark, Norway and Luxembourg thought that the special difficulties which existed in their countries rendered it necessary for their Governments to give the proposals a more detailed examination before giving their decision.

The accord remained open for signature by other members of the United Nations and by neutral countries until 1 January 1947.

Chapter Nine
Overcast and Paperclip

As territory was occupied after the European landings, both USAAF and US Navy NAVTECHMISEU teams roamed far and wide, sometimes just behind the advancing troops, questioning, searching, and trying to find the answers to Germany's amazing wartime technical progress. One day in April 1945, while one of the Navy teams was searching at Oberamniergau in Bavaria, they found a group of German missile designers and their leader, Professor Herbert Wagner. He had been the chief missile design engineer for the Henschel aircraft works and had masterminded the development of the Hs-293, a radio-controlled glide bomb. In the nearby Hartz Mountains, buried blueprints, models, and prototypes were found.

By early May 1945, Professor Wagner, his four assistants and their files were in Washington. Many organizations were interested in exploiting them, including the Navy Bureaus of Aeronautics and Ordnance and the Army Air Force, but none was willing to take custody of the missile team. So they were placed in a Washington hotel, where officers from the Office of Research and Inventions - later the Office of Naval Research - stood watch as Wagner and his men worked to perfect a controlled anti-aircraft rocket for use in the continuing war against the Japanese.

Wagner first worked at the Special Devices Center, at Castle Gould and at Hempstead House, Long Island, New York; in 1947, he moved to Naval Air Station Point Mugu.

The hotel arrangement was too expensive for ONI's funding resources. so the National Advisory Committee for Aeronautics were asked to help.

A captured German V-2 rocket is moved by an American Army transporter. Many V-2s ended up being launched by Nazis moved to the USA under Operation Paperclip. (USAAF)

Wernher von Braun - seen here with his arm in a cast following a car accident - surrendered to the Americans just before this 3 May 1945 photograph was taken. The American high command was well aware of how important their catch was: von Braun had been at the top of the Black List, the code name for the list of German scientists and engineers targeted for immediate interrogation by U.S. military experts.

What was needed was a secluded estate where life would be pleasant but secure. The Guggenheim Foundation was found to have such a place, the Jay Gould medieval castle at Sands Point on Long Island, which became the Special Devices Center of the Office of Naval Research. Initially its use was kept quite secret; guards were placed at the gate, and no Germans left the grounds except under escort.

In the summer of 1945, the Technical Information Center published German Technical Aid to Japan to delineate '...*those German techniques, devices and weapons, the use of which by the Japanese would have a bearing on the war in the Pacific*'.

The surrender of the submarine U-234 to US forces at the time of Germany's collapse contributed significantly to the survey: the submarine had been *en route* to Japan with a valuable cargo, including complete drawings for the Messerschmitt Me-163 rocket fighter, an entire German electronics library, fire-control equipment, radar, and radio equipment.

Dr. Heinz Schlicke, a German electronics expert, was one of the passengers aboard the U-234. He was going to deliver a series of lectures in Japan on German electronic development and had extensive documentary material with him. Arrangements were made for Dr. Schlicke to give the same lectures in the Navy Department between 19 and 31 July 1945.

To ensure a more widespread distribution of the information from Dr. Schlicke's lectures and documents, the Technical Intelligence Center issued 800 copies of *Electronics Research in the German Navy* on 15 September 1945. The book covered a general review of electronic research in the German navy, U-boat camouflage, radar and search receivers on U-boats, infrared in the German navy, communications with submerged U-boats over great distances, direction-finding in U-boat warfare, ships' antennae, prevention of radio transmitted direction-finding and some observations on German techniques in the use of centimetre waves and the theory of line transformers.

The Technical Intelligence Center also edited and distributed 12,000 copies of German Admiral Doenitz's essay, '*The Conduct of the War at Sea*', a review of the German navy's participation in World War Two'. The surrender of

U-234 following VE Day brought more scientists to the Special Devices Center. They included Dr. Falck, one of the German navy's top ship designers; the head production engineer for the Messerschmitt works; and several experts on night fighter techniques.

From the information developed by the interrogation of the Wagner Group, Dr. Schlicke, and others, the Navy realized that a most valuable reparation from Germany could be the brains of its scientists. The various technical bureaus of the Navy began to show more interest in acquiring the services of some of' the German technical specialists. As the Army and Army Air Force were also interested in procuring German specialists, it was determined that only a joint programme under the general administration of the Joint Chiefs of Staff (JCS) could be effective for successful and reasonable exploitation.

The Joint Intelligence Committee (JIC) of the JCS served as the intelligence arm, responsible for advising the JCS on the intelligence problems and policies and providing intelligence information to the JCS and the Department of State. The JIC was composed of the Army's director of intelligence, the chief of naval intelligence, the assistant chief of Air Staff-2, and a representative of the Department of State. Oh how the military - especially the US Navy - love their acronyms!

In July 1945, the Secretary of State and the Secretary of War agreed to establish Project Overcast, a German scientist procurement and exploitation programme initially in the rocket and guided-missile research area. The first scientists to arrive under the project came on 8 July 1945 for short-term interrogation and exploitation by the Army, Navy, and Air Force. An additional twenty-two scientists were taken to the United States in September. One stated purpose of Operation Overcast was to deny German scientific expertise and knowledge to the Soviet Union and the United Kingdom, as well as inhibiting post-war Germany from redeveloping its military research capabilities. So much for co-operation between Allies!

On 27 August, the JIC recommended that the Joint Intelligence Objective Agency (JIOA) be authorized to set up an interim procedure to coordinate the temporary exploitation of German and Austrian specialists, scientists, and technicians in the United States pending the formation of approved government exploitation policy and procedures. On 12 September, the JCS, with the concurrance of the State-War-Navy Co-ordinating Committee (SWNCC), approved the interim procedures to be used by the JIC through the JIOA.

Thus came into being an operation overseen by the Office of Strategic Services (OSS) – forerunner to the Central Inteligence Agency (CIA) in which over 1,500 German scientists, technicians, and engineers from Nazi Germany and other foreign countries were taken to the United States for employment.

The JIOA was given direct responsibility for operating the foreign scientist programme, initially code-named Overcast and subsequently Paperclip. The JIOA was composed of one representative of each member agency of the JIC, and an operational staff of military intelligence officers from the different military services. Among the JIOA's duties were administering the programme's policies and procedures, compiling dossiers, and serving as liaison to British intelligence officers operating a similar project. It was also responsible for collecting, declassifying, and distributing Combined Intelligence Objectives Subcommittee (CIOS) and other technical intelligence reports on German science and industry. In addition, the JIOA took over many of the activities of CIOS when that organisation was terminated. The JIOA was disbanded in 1962.

Although the JIOA's recruitment of German scientists began after the Allied victory in Europe on 8 May 1945, President Harry Truman did not formally order the execution of Operation Overcast until August 1945. Truman's order expressly excluded anyone found '*...to have been a member of the Nazi Party, and more than a nominal participant in its activities, or an active supporter of Nazi militarism*'. However, those restrictions would have rendered ineligible most of the leading scientists the JIOA had identified for recruitment, among them rocket scientists Wernher von Braun, Kurt H. Debus and Arthur Rudolph, and the physician Hubertus Strughold, each each of whom had earlier been classified as a '*...menace to the security of the Allied Forces*'.

To circumvent President Truman's anti-Nazi order and the Allied Potsdam and Yalta agreements, the JIOA worked independently and secretively to create false employment and political biographies for the scientists. The JIOA also expunged from the public record the scientists' Nazi Party memberships and régime affiliations. Once 'sanitised' or 'sheep-dipped' of all vestiges of their Nazism, the scientists were granted security clearances by the U.S. government to work in the United States.

The project code name was changed from Overcast to Paperclip on 10 November 1945, following the compromise of the former code name. By the end of 1945, 132 scientists had been brought to the United States under Project Paperclip. Allegedly, 'Paperclip', the project's operational name, derived from the paperclips used to attach the scientists' new political personae to their 'US Government Scientist' JIOA personnel files.

The policy and procedures for exploitation of German and Austrian scientists in the United States were submitted by SWNCC to JCS for comment on 26 February 1946 and on 4 March it was approved by SWNCC and sent to the JCS for execution by JIOA.

By April 1946, approximately 155 German scientists and technologists were in the United States under Project Paperclip for exploitation by the military services, all under voluntary contracts not to exceed twelve months. By the end of July, the figure had risen to 190 scientists, and there were over 200 others whose services had been requested by the various technical services of the War and Navy Departments.

Rocket Scientists and the Osenburg List
Having failed to conquer the USSR with Operation Barbarossa (June–December 1941), the Siege of Leningrad (September 1941 – January 1944), Operation Nordlicht ('Northern Light', August–October 1942), and the Battle of Stalingrad (July 1942 – February 1943), Nazi Germany found itself at a logistical disadvantage. The failed conquest had depleted German resources and its military-industrial complex was unprepared to defend the *Großdeutsches Reich* (Greater German Reich) against the Red Army's westward counterattack. By early 1943, the German government began recalling from combat a number of scientists, engineers, and technicians; they returned to work in research and development to bolster German defence for a protracted war with the USSR. The recall from frontline combat included 4,000 rocketeers returned to Peenemünde, in northeast coastal Germany.

Overnight, Ph.D.s were liberated from menial duties, masters of science were recalled from orderly service, mathematicians were hauled out of bakeries, and precision mechanics ceased to be truck drivers.

The Nazi government's recall of their now-useful intellectuals for scientific work first required identifying and locating the scientists, engineers, and technicians, then ascertaining their political and ideological reliability. Werner Osenberg, the engineer-scientist heading the *Wehrforschungsgemeinschaft* (Military Research Association), recorded the names of the politically-cleared men, thus reinstating them to scientific work.

This became known as the Osenberg List, and in March 1945 at Bonn University, a Polish laboratory technician found pieces of the list stuffed in a toilet; the list subsequently reached MI6 in the United Kingdom, which onward-transmitted it to U.S. Intelligence. Then U.S. Army Major Robert B. Staver, Chief of the Jet Propulsion Section of the Research and Intelligence Branch of the U.S. Army Ordnance Corps, used the Osenberg List to compile his list of German scientists to be captured and interrogated; Wernher von Braun, Nazi Germany's premier rocket scientist, headed Major Staver's list.

Staver's original intent was only to interview the scientists, but what he learned changed the operation's purpose. On 22 May 1945, he transmitted to U.S. Pentagon headquarters Colonel Joel Holmes's telegram urging the evacuation of German scientists and their families, as most 'important for [the] Pacific war' effort. Most of the Osenberg List engineers worked at the Baltic coast German Army Research Centre Peenemünde, developing the V-2 rocket. After capturing them, the Allies initially housed them and their families in Landshut in southern Germany.

Allied Intelligence described nuclear physicist Werner Heisenberg, the German nuclear energy project principal, as *'...worth more to us than ten divisions of Germans.'* In addition to rocketeers and nuclear physicists, the Allies also sought chemists, physicians, and naval weaponeers.

Meanwhile, the Technical Director of the German Army Rocket Centre, Wernher von Braun, who had surrendered to the Americans on 3 May 1945, was taken along with his department chiefs from Garmish to Munich on 19 June 1945, two days before the scheduled handover of the area to the Soviets, by US Army Major Robert B. Staver, Chief of the Jet Propulsion Section of the Research and Intelligence Branch of the U.S. Army Ordnance Corps in London, and Lt Col R. L. Williams. The group was flown to Nordhausen, and was evacuated 40 miles southwest to Witzenhausen, a small town in the American Zone, the next day. Von Braun was briefly detained at the 'Dustbin' interrogation centre at Kransberg Castle where the elite of the Third Reich's economy, science and technology were debriefed by U.S. and British intelligence officials. There is some evidence to suggest that British intelligence and scientists were the first to interview him in depth, eager to gain information that they knew US officials would deny them.

On 20 June 1945, the U.S. Secretary of State approved the transfer of von Braun and his specialists to America; however, this was not announced to the public until 1 October 1945. Von Braun was jailed at P.O. Box 1142, a secret military-intelligence prison in Fort Hunt, Virginia. Since the prison was unknown to the international community, its operation by the US was in violation of the Geneva Convention of 1929, which the United States had ratified. Although Von Braun's interrogators pressured him, he was not tortured; however, another PoW, U-boat Captain Werner Henke was shot and killed while climbing the fence at Fort Hunt.

In August 1945, Colonel Holger Toftoy, head of the Rocket Branch of the Research and Development Division of the U.S. Army's Ordnance Corps, offered initial one-year contracts to the rocket scientists; 127 of them accepted. Officially, the first seven rocket technicians arrived in the United States at New Castle Army Air Field, just south of Wilmington, Delaware, on 20 September 1945 to be flown to Boston and then taken by boat to Fort Strong, located on Long Island in the middle of Boston harbour: these were Wernher von Braun, Erich W. Neubert, Theodor A. Poppel, August Schulze, Eberhard Rees, Wilhelm Jungert, and Walter Schwidetzky.

Early on, the United States created the Combined Intelligence Objectives Subcommittee (CIOS). This provided the information on targets for the T-Forces that went in and targeted scientific, military and industrial installations (and their employees) for their know-how. Initial priorities were advanced technology, such as infrared, that could be used in the war against Japan; finding out what technology had been passed on to Japan; and finally to halt the research. A project to halt the research was codenamed 'Project Safehaven', and it was not initially targeted against the Soviet Union; rather the concern was that German scientists might emigrate and continue their research in countries such as Spain, Argentina or Egypt, all of which had sympathised with Nazi Germany.

Much U.S. effort was focused on Saxony and Thuringia, which by 1 July 1945 would become part of the Soviet Occupation zone. Many German research facilities and personnel had been evacuated to these states, particularly from the Berlin area. Fearing that the Soviet takeover would limit U.S. ability to exploit German scientific and technical expertise, and not wanting the Soviet Union to benefit from the same expertise, the United States instigated an 'evacuation operation' of scientific personnel from Saxony and Thuringia, issuing orders such as:

'On orders of Military Government you are to report with your family and baggage as much as you can carry tomorrow noon at 1300 hours (Friday, 22 June 1945) at the town square in Bitterfeld. There is no need to bring winter clothing. Easily carried possessions, such as family documents, jewelry, and the like should be taken along. You will be transported by motor vehicle to the nearest railway station. From there you will travel on to the West. Please tell the bearer of this letter how large your family is.

By February 1946, over 100 Operation Paperclip scientists had arrived at Fort Bliss to develop rockets and were attached to the Office of the Chief of Ordnance Corps, Research and Development Service, Suboffice (Rocket), headed by Major James P. Hamill. Although the scientists were, according to the records, initially '...*pretty much kept on ice*' a phrase that resulted in the unofficial nickname Operation Icebox, they were subsequently divided into a research group and a group who assisted with V-2 test launches at White Sands Proving Grounds. German families began arriving in December 1946, and by the spring of 1948, the number of German rocket specialists nicknamed locally as 'Prisoners of Peace' in the US was 127. In 1953, funding cuts caused the cancellation of work on the Hermes B2 ramjet work that had begun at Fort Bliss.

A revision of Paperclip was considered urgent. The morale of the first scientists to arrive was low because of the failure to implement an approved policy under which their families could join them in the United States. The 350-scientist ceiling needed to be raised to 1,000 in order to meet the demands of both the technical services and civil research programs.

The top salary of $10.00 per day paid to dependents in German marks had become insufficient inducement for highly qualified German scientists, who were being offered three to five times that amount by the French and Russians in addition to being permitted to take their families with them.

The revision was coordinated in the War Department General Staff and was approved by the Secretary of War on 31 July 1946. The resulting State War Navy Coordinating Committee (SWNCC) policy was recommended to the President by Secretary of State James F. Byrnes in August and was approved by President Truman on 3 September 1946. In spite of White House approval, there was continuing obstruction to Paperclip by elements in the State Department. Samuel Klaus was the State Department representative on the committee set up to formulate a plan to implement the SWNCC policy. Capt. Bosquet N. Wev, head of JIOA, was chairman of the committee. Klaus, whose primary interest was in getting displaced persons from Germany into the United States, looked upon the project as depriving him of immigration quotas. Furthermore, he considered any German scientist who had performed any service in support of the German war effort as having accordingly performed services detrimental to the US war effort and based on the War Powers Act, was to be considered prejudicial to the best interests of the United States and thus not eligible for immigration. It apparently made no difference to Klaus that bringing German scientists to the United States had been approved by Secretary of State Byrnes and President Truman.

As a result of Klaus's actions, many of the 1,500 German and Austrian scientists that the U.S. Army and Navy had wanted to bring to the United States were taken over by the Soviet Union for exploitation of their knowledge and experience. The military technical bureaus and laboratories that could have used these highly qualified people kept putting pressure on JIOA for action, and JIOA kept trying to pass the word up to the top levels of the State Department to get around the obstructionism there. But each channel seemed to run into someone officially subservient to Klaus.

By 10 January 1947, there were still only 285 Paperclip scientists in the United States, and 390 others had been requested for employment. The German scientists working in the United States were either under military employment or working on military projects in industries that had contracts with the War or Navy Departments. Finally, Capt. Wev, through his own efforts, arranged for a hearing before a Senate appropriations subcommittee in June 1947. As a result of his testimony, appropriations to the State Department were blocked by the Senate until the barrier to Project Paperclip was broken.

As of late 1949, eighty-two German scientists were still employed by the US Navy in Project Paperclip. They were located at a number of naval stations on both coasts. Others were located in industrial cities and were accountable to inspectors of naval material and branch offices of the Office of Naval Research. The Bureau of Medicine had some Germans working on medical research at Bethesda and at the Submarine Base, New London.

German scientists were used in a variety of fields. For example, the Naval Ordnance Laboratory at White Oak, Maryland, employed them in three main areas: designing facilities for research in acoustic, supersonic, and explosive phenomena; conducting basic research in physical optics, fluid mechnanics, acoustics, and explosives; and designing specific weapons and weapons countermeasures.

A group of 104 rocket scientists, many former German Nazis, at Fort Bliss, Texas. They had moved there under Operation Paperclip.

The Soviet Union had competing extraction programmes known as trophy brigades operating under Operation Osoaviakhim.

Many Project Paperclip personnel working for Bureau of Ordnance were experts in aerodynamics and had had extensive previous experience with the operation of the Kochel supersonic wind tunnel that had been brought to the United States and installed at the laboratory at a cost of $2.5 million. By late 1945, it was becoming obvious that to meet new needs the ONI Technical Intelligence Section needed greater numbers of more diversified technical experts. As a result, additional civilian experts were hired in nuclear energy, aeronautics, naval engineering, and other fields.

One of the first key civilians to be employed was Dr. A. Keith Brewer, a physical chemist who had specialised in isotope operation processes at the Bureau of Standards and who came to ONI in 1946. Brewer was a strong believer in what he termed as 'the scientific method' and pushed for more and better scientifically trained personnel in the intelligence production process. Like all true scientists, he was a sceptic and for a decade urged proof positive in appraising Soviet and Communist Chinese nuclear developments.

Another early civilian employee whose experience had been more in applied sciences was Dr. Jack Alberti, who had served in the Special Activities Branch of ONI in the prisoner-of-war interrogation field throughout World War Two. His specialty was captured enemy equipment and scientists.

William E. W 'Bill' Howe, an electronics engineer, came in February 1949 from the Naval Research Laboratory, where he had been involved in testing captured Japanese equipment and in developing a receiver to intercept German missile guidance signals. At ONI, Howe was soon involved in electronic reconnaissance activities.

Mr. R. E. Kinzy, an aeronautical engineer, came from the Glenn L. Martin Company to do some of ONI's earliest Soviet missile threat analysis.

In 1946, the United States Bureau of Mines employed seven German synthetic fuel scientists at a Fischer-Tropsch chemical plant in Louisiana, Missouri. And so the list went on.

In early 1950, legal U.S. residency for some of the Project Paperclip specialists was brought about through the U.S. consulate in Ciudad Juárez, Chihuahua, Mexico; thus, Nazi scientists legally entered the United States from Latin America.

Eighty-six aeronautical engineers were transferred to Wright Field,

14 April 1955 and in an unprecedented ceremony held at Huntsville High School, one hundred and three German-born scientists, technicians, and members of their families became American citizens in a mass ceremony. Among those taking the oath of citizenship was Dr. Wernher von Braun. This had been preceded on 11 November 1954 when thirty-nine German-born scientists, along with the wives of two of the Operation Paperclip group, were sworn in as U.S. citizens. This event had a favourable impact on the Redstone Arsenal's overall public relations image because it eliminated the stigma of 'aliens' working on highly classified missions for U.S. defence.

where the United States had Luftwaffe aircraft and equipment captured under Operation Lusty.

The United States Army Signal Corps employed twenty four specialists – including physicists Georg Goubau, Gunter Guttwein, Georg Hass, Horst Kedesdy, and Kurt Lehovec; physical chemists Rudolf Brill, Ernst Baars, and Eberhard Both; geophysicist Helmut Weickmann; optician Gerhard Schwesinger; and engineers Eduard Gerber, Richard Guenther, and Hans Ziegler.

In 1959, 94 Operation Paperclip men went to the United States, including Friedwardt Winterberg and Friedrich Wigand. Throughout its operations to 1990, Operation Paperclip imported 1,600 people as part of the intellectual reparations owed to the United States and the UK, some $10 billion in patents and industrial processes.

During the decades after they were included in Operation Paperclip, some scientists were investigated because of their activities during World War Two when their previous activities started to surface. Arthur Rudolph was deported in 1984, but not prosecuted, and West Germany granted him citizenship. Similarly, Georg Rickhey, who came to the United States under Operation Paperclip in 1946, was returned to Germany to appear at the Dora Trial in 1947; he was acquitted, and returned to the United States in 1948, eventually becoming a U.S. citizen. The aeromedical library at Brooks Air Force Base in San Antonio, Texas, had been named after Hubertus Strughold in 1977. However, it was later renamed because documents from the Nuremberg War Crimes Tribunal linked Strughold to medical experiments in which inmates from Dachau were tortured and killed.

There were many others.

No one has done more for aerodynamics than Dr. Ludwig Prandtl from the University of Göttingen, Germany. He was born in 1874 and lived to 1953. Motivated by the flights and research of the Lilianthal brothers and the Wrights, Prandtl formulated a large number of major theories which were and are still being used today to design practical aircraft.

Prandtl stated that Dr. Hans Multhopp was his best student. Multhopp designed the P.183 and is famous for his T-tail designs. After the war he first worked in England and then in 1949 he went to the US under Paperclip and worked for the Martin Aircraft Co. He went on to design a large number of aircraft. Much of the design of the Martin XB-51 was performed by Multhopp working as a consultant.

Dr. Adolph Busemann is famous for his work in developing the swept wing theory in the 1930's. He presented his paper on swept wings at the 5th Volta Conference in Rome, Italy. Many people, including von Kármán, made fun of his swept wing but all swept wings of today can trace their lineage to the original data from Busemann. In 1946, he went to the US under Operation Paperclip and worked at NACA Langley before in 1950 he became a professor at the University of Colorado. Busemann also built the first supersonic wind tunnels in Germany in 1939.

Dr. Woldemar Voigt designed the Me.262 and P.1101 jet aircraft at Messerschmidt. Before the war's end he had redesigned the Me.262 by placing the turbines in the wing roots. Voigt's Me P.1107/B, heavy bomber design was used in the almost identical design of the British Vickers Valiant Mk. 1 bomber in 1950. In 1946 he moved to the US under Paperclip and worked for Bell Aircraft on the X-5 and other aircraft.

Dr. Richard Vogt was known for his wild aircraft designs. One such design, the P.202, designed in 1944, was called his "Swivel Wing" which was later picked up by R.T. Jones at NASA Ames and built as the AD-1 "Oblique Wing" in 1980. After graduating from the University of Stuttgart he joined the Dornier Co. in 1923. Later he designed a large number of aircraft for

A V-2 test launch at the White Sands Missile Range. The foreshortening effect of the photgraphers telephoto lens make the observers look closer to the rocket than they really are!

Blohm & Voss. After the war he came to the US under Paperclip. Working at the Boeing Aircraft Co., he continued to design unusual aircraft such as nuclear-powered bombers in the mid 1950s.

Dr. Ing. Anselm Franz, director of the gas turbine engine development at Junkers in Germany from 1940 to 1945, came to the US with Operation Paperclip in 1946 and became vice president and assistant general director at AVCO Corp. Lycoming Division in Connecticut where he continued his management efforts developing gas turbine engines. Most Lycoming gas turbine engines were designed by Franz and his team of engineers.

The full scope of Operation Paperclip and its effects on the western world may never be fully realised. What started as an attempt to maximise the potential of captured German materiel and expertise became a series of tools to exploit the commercial potential for American industry and to be used against the Soviet Union, their former allies.

Chapter Ten
Across the Channel

It was not just the British and the Americans who were interested in testing captured German equipment. The French made use of three Messerschmitt Bf109Es captured prior to the occupation of France in June 1940, one of which was destroyed in a mid-air collision with a Morane MS406 of the French Armée de l'Air within a few weeks of capture in 1939. A second example was flown at Orleans/Bricy and then handed over to the RAF in May 1940, receiving the serial number AE479. The third example, WNr.3326 'Red 9' of I/JG 51, was also flown to England, but was not flown after its arrival there. It originally force-landed near Nancy on 20 September 1939 and after testing by the French was flown to Tangmere on 22 December 1939.

Following the occupation of France, the Free French Air Force in North Africa took over a number of Axis aircraft. These included an Italian CANT Z1007 bis Alcione, captured at Damascus in Syria and later used by the Lignes Aeriennes Militaires (LAM), the French military air transport organisation in North Africa. One, or possibly two, Junkers Ju.52/3ms were restored to airworthiness after being found in Tunisia in 1943. These examples were probably later used in mainland France.

Many ex-Luftwaffe aircraft were transferred to France for operational use by the post-war Armée de l'Air. These aircraft were, in the main, serviced by RAF Repair and Salvage Units at Leck and Lüneburg, or by French units attached to them. When serviceable, the aircraft were delivered to France. A large number of other unserviceable aircraft requiring repair, or for use as spares, were sent by rail from Leck to an Armée de l'Air depot at Nanterre, near Paris. These included all the jet, rocket and other aircraft selected for evaluation by the French at such test centres as the Centre d'Essais en Vol (CEV) at Melun/Villaroche, or at the Centre d'Experimentation Aeriennes Militaires (CEAM) at Mont-de-Marsan in the post-war period.

The Dornier Do.335 *Pfeil* (Arrow) with its engines mounted at the front and rear of the fuselage driving both a tractor and a pusher propeller was of particular interest to the French, for they had been developing an aircraft of similar design, and so the idea surfaced to compare the two configurations.

The French Arsenal VB 10 was a low-wing monoplane with retractable tailwheel undercarriage and of largely orthodox configuration. The ultimate product of a design that began with the Arsenal VG 10 prior to the war, the VB 10 added a second engine behind the cockpit which drove a second propeller, coaxial with and counter-rotating to the propeller driven by the engine in the nose.

The 5th Bureau of the General Headquarters of l'Armée de l'Air (EMGAA) issued a list of German equipment on 10 February 1945 which could be of use to the French Air Force at the end of the war. Among a selection of light and medium bombers was the Do.335.

The first *Pfeil* acquired by the French was captured on 27 April 1945, by the 1st DB (lst Armoured Division) when it overran the test airfield at Mengen, in southern Germany. This was Do.335 M14, Werk.nr 230014, RP+UQ. This was the second development aircraft for the Do.335B-2, a heavily armed day fighter or *Zerstörer*. Allegedly it had first flew in November 1944, and a close inspection revealed that it should be possible to restore it to flying condition. The plan was to dismantle it for road transport into France, after which it would be reassembled and flown to the Paris area and handed over to the Société Nationale de Construction Aéronautique du Sud-Ouest (SNCASO), together with a recovered Messerschmitt Me.262. Both aircraft would then be overhauled and tested by the Centre d'Essais en Vol (CEV).

Colonel Badré, previously a test pilot with the pre-war Centre d'Essais des Materiels Aériens (CEMA – the predecessor of the CEV), was in command of the 5th Bureau, and he instructed Adjutant-Chef (Warrant Officer) Fostier to proceed to the airfield to familiarise himself with the aircraft and supervise its dismantling. The plan was to transfer the machine to Lyon-Bron airfield, the home of Arsenal de l'Aéronautique.

Badré made sure that the Dornier would be in the hands of mechanics familiar with twin-engined, tricycle-undercarriaged aircraft. So it was that Compagnie de Ravitaillement et de Reparation Technique 83 (CRRT 83 - Supply and Repair Company) was chosen. Capitaine Dispot gathered a team of technicians and hurried to Mengen, which was not at all safe, as the mine disposal team with the 1st DB had only cleared a single path to the small dispersal where the Do.335 stood.

Crates of special tools for the Do.335 were located on the airfield, and three German engineers specialising in hydraulics, electrics and weapons were conscripted, simply by being 'invited' along with their families to settle for a while in Mengen.

Lack of technical information was a challenge to the mechanics, who had to explore the technology of the aircraft from scratch, a difficult task made even more difficult by the machine being a prototype. A number of Dornier blueprints were discovered in a house in Mengen – around the same time, the French authorities had made contact with Claudius Dornier and his son to elicit more information.

Back at the dispersal, M14 had problems with an unreliable locking mechanism on its nose leg. The team made its way to Ravensburg where, in an underground workshop and in a large hangar, several apparently completed Do.335s had been discovered, although their mainspars had been sabotaged. A complete undercarriage assembly was located, and several ancillaries.

While this was going on, the EZ 42 gyroscopic sight was dismantled from the aircraft at Mengen and quickly sent to Paris.

The French Arsenal VB 10 seen post war with some of the building team. The aircraft name use of some Dornier 335 technology.

Do.335B-2 M14 RP+UQ as captured by French soldiers at Mengen, Germany on 27 April 1945.

The whole area was in considerable chaos, as the occupation zones were not then clearly defined, so the order was issued to have the Dornier transported to Lyon by road as quickly as possible in case the Russians appeared in that sector.

The Do.335 was pulled into a hangar, quickly dismantled and despatched on three trailers. As it was a prototype, the hydraulic lines had few connectors and the electrical circuits, similarly, bore few plug/socket connections. It was necessary to cut many before removing the wings, the severed ends being carefully labelled to enable reconnection.

Having crossed the Rhine over a pontoon bridge, the convoy reached Lyon-Bron and work was immediately begun to restore the aircraft to flying condition. A friendly hand was given by teams from Bron-based Marauder units, and the Do.335 was partly Americanised, thanks to the use of American standard hydraulic and electrical connectors, these being far from compatible with metric DIN equipment. The two MK 103 cannon were replaced by 150kg of ballast in order to maintain the centre of gravity. Without this, the fighter tipped on to its tail. Lack of documentation prevented any testing of the weapons.

Reassembly began in July but took longer than expected. Engine tuning proved difficult due to the type of spark plugs and the octane rating of the fuel. Eventually, ground runs using American 100/130-octane fuel and German oil proved satisfactory. Owing to the lack of time, the DB.603s had not been inspected in any depth, and the team had to rely on these ground runs to assess their condition.

The Do.335 was fitted with an ejection seat. Before abandoning the aircraft the pilot manually jettisoned the canopy, then set off explosive charges at the root of the upper fin and within the rear spinner by depressing an electrical switch to his right, protected by a red cover. This removed the upper fin and rear propeller, thus giving the pilot a clear exit when baling out. He then operated the ejection system, which employed compressed air pressurised at 120kg/cm^2. The subsequent acceleration peaked at about 18G! If he survived the ejection, the pilot had to jettison the seat and open his parachute.

During overhaul the ejection device was removed, but, as the team had several others in store, it was decided to test the power of the jack. It was attached to an eight millimetre diameter cable anchored to the hangar floor. One of the mechanics activated the system by means of a long tool, whereupon the piston severed the cable and made a hole in the concrete roof of the hangar. This trial was watched by a rather worried future pilot! Later, when the seat was

There were a number of abandoned Do.335s around that could be used for spares.

reinstalled, the team made a rather unpleasant discovery. The leather-cloth headrest was stuffed not only with horsehair, but with human hair as well. The padding was entirely replaced.

By the end of July the first ground run was accomplished in front of the control tower at Lyon-Bron, with Jean Girault at the controls. Shortly after, however, the team was ordered to Chartres for a few days and, during this time, personnel from Arsenal took the opportunity to study the Dornier. They also repainted it using the same light grey paint with which they had finished the Arsenal VB-10-01. These Frenchmen took exception to Capitaine Dispot leaving German markings on the aircraft.

On 5 August the aircraft was declared fit to fly. Colonel Badré arrived two days later and instructed that the undercarriage and flaps of the Dornier be checked before he made two taxi runs, one at slow speed on the perimeter track, and the other at higher speed on the main runway. These trials revealed that, with an ambient temperature of about 20°C, the engines warmed up rapidly, even without a run-up. It was clear that the engines were not adequately cooled on the ground, so the taxying trials were discontinued. Moreover, the brakes were inefficient, and considerable effort was needed by the pilot on the toe brakes to produce a weak deceleration.

The first flight took place on 8 August with Paul Badré at the controls. His report reads as follows: *"That morning, the weather conditions were as follows. Slight northerly wind, sky completely overcast with stratocumulus, base approximately 1,000 metres and visibility of 10 to 20 kilometres. I decided to do the first flight. Take-off was normal after a roll of about 1,000 metres, and at an indicated speed of 200 kilometres per hour, with 2,600 rpm and 1-3ata [atmospheres pressure] of boost on each engine. The aircraft rolled absolutely straight, with no tendency to swing to either side.*

Retraction of undercarriage and flaps was very good. Climb to 1,000m was accomplished at 300km/h and 1-2ata. At 1,000m I levelled the aircraft and set ata boost/2,000 r.p.m. to achieve a speed of 400km/h. At this stage the coolant and oil temperatures were 75° and 50°C respectively for both engines, oil pressure showing about 7kg/cm^2. After a few manoeuvres around the airfield to check the controls and

Do.335 'No.1' at Brétigny in 1947.

trim, which seemed perfect, I prepared to make a last fly-past before selecting wheels down.

Suddenly the aircraft started vibrating violently and the cockpit filled with smoke. As I feared damage to the extended propeller shaft, I immediately stopped the rear engine. Seeing no improvement, I did the same to the front engine. Then the smoke decreased, and I discovered that the airspeed indicator was not working, probably because its probe had been damaged by vibration. I lowered the flaps and undercarriage and, after two 'S' turns, landed on the main runway pointing north. The landing was rather firm, but the shock absorbers did not compress by more than half their stroke.

The right tyre burst, and, as the brakes were dead, the aircraft left the runway and crashed into a belly-landed B-26. A strut of the starboard main undercarriage broke and the aircraft came to rest on its right wingtip'.

The aircraft made a hard landing, bounced once and then entered a high-speed groundloop. As both engines were stopped, the extension of flaps and undercarriage had drained residual hydraulic power, leaving nothing for brakes. Without these, the machine left the runway and the right wheel encountered a small stack of bags of cement, destroying sheets of corrugated iron which covered it. The aircraft continued at about 100 km/h, slicing in two the fuselage of the crashed Marauder before coming to a halt 50 metres beyond, settling down on its right wingtip as the right main leg retracted.

Badré escaped unscathed from an almost undamaged aircraft which, seconds before, he had seen rushing out of control towards an abandoned bomber. Inspection revealed a serious oil leak on the front engine, caused by a fractured connecting rod penetrating the crankcase just below the attachment point of one of the bearers. The smoke in the cockpit had probably come from oil spraying on the exhaust stubs. The fuselage was only slightly damaged, but the lower fin had been demolished and the starboard wing and undercarriage, together with the forward engine, needed to be replaced or repaired.

Investigation revealed that a nut on the connecting rod had not been properly secured, perhaps through an act of sabotage.

On 19 September Headquarters ordered that the prototype be transferred by road to SNCASO, and this was accomplished with the help of Secteur de l'Air No 3 (Air Sector No 3). The Do.335 was repaired in Suresnes, at 40 Rue Jean-Jacques Rousseau, and repainted once again, this time in green.

Following the repairs after its landing crash 'No 1' was sent to the Centre d'Essais en Vol (CEV) at Brétigny on 3 June 1946. There it was checked and weighed, and the CEV took the opportunity to perfect the adjustment of several systems. Between 12 and 19 February 1947, an initial modification programme was performed, and it was only then that a parachute was refitted to the pilot's

Right: A poor quality picture of Roger Receveau in the cockpit of Do.335 No.2. Note the hatch with glazing aft of the pilot's cockpit for the radar operator.

Below: the radar operators position in No.2, with the German radar equipment removed and replaced with a seat, oxygen supply and a flight recorder box.

seat. As the rearward centre of gravity limit was not clearly identified, the guns were reinstalled, bringing the centre of gravity forward, and 200kg (440lb) of ballast was added to the machine to offset the lack of ammunition.

Roger Receveau, who was involved with testing the tandem-engined Arsenal VB-10, took over the trials. He started with a series of taxying runs on 24 February. He experimented with runs using only the rear engine, starting with low-speed taxying and made ground-runs of the engines against the brakes before high-speed trials. The brakes and wheels overheated rapidly, and had to be cooled using a water hose. That same day a wheel caught fire after an emergency braking test. No real damage was done, and the aircraft was brought back to the hangar on the front engine.

The first flight in the hands of Receveau took place on 13 March 1947, to check general handling and establish minimum control speed in the air. He tested the controls at different speeds and in several configurations, and also dived to an indicated airspeed of 600km per hour. On 25 April, 8 May and 14 May Receveau took the single-seat Do.335 No 1 up on three further familiarisation flights, checking the general behaviour of the airframe and engines. These flights were intended to examine longitudinal stability rather than overall performance. However, the second flight saw speed increased to an impressive 700km per hour, with stabilised level flight producing 650-660km per hour.

During this series of tests the effect of controls at different speeds and in various configurations was measured. The aircraft was rather heavy to handle at low speeds, but better at medium to high speed, with good control response. The engines, however, were found to be difficult to manage and some instruments, such as the temperature gauge and compass, were useless because of their erratic behaviour. Despite these relatively minor problems, Receveau flew the aircraft on one engine on the third flight. Also, its 200kg of ballast was removed, moving the centre of gravity rearwards from 40.3 per cent to 41.5 per cent mean aerodynamic chord. On completion of these four flights the aircraft was returned to the workshop to have a punctured fuel tank repaired. Refitting the tank, and replacing the internal fuel pump, required the addition of a small fairing atop the fuselage, the work not being completed until November that year.

A second aircraft was also overhauled and repaired at Mengen by the French and German teams. This was Do.335 M17 Werk.Nr 240313, the development aircraft for the Do.335B-6 two-seat night-fighter, which had been discovered in an incomplete state. A radar operator was accommodated in a small cockpit midway along the fuselage, his entry and exit being via a small glazed hatch.

M17 was flight-tested at Mengen by Roger Receveau on 2 April 1947. It was a trouble-free flight, and both teams enjoyed a drink under the aircraft afterwards, in a relaxed atmosphere. However, two days later, during a ground run, the drive-shaft of the supercharger on the rear engine failed and had to he replaced. Following this, the same drive gear on the front engine failed as soon as the engine was started. In this case a replacement DB 603E was installed with a special declutch mechanism for the supercharger fitted.

The Do.335 was then able to resume flying, and was air-tested for half an hour on 29 May. That same day it was ferried to Brétigny in the hands of Receveau, the flight taking 80 minutes.

The two Do.335s remained on the ground until October 1947. Then M17 resumed flying on 20 October, and N114 performed its sixth test a month later, on 20 November. Testing alternated between the two machines, but at a very slow rate. The second seat in M17 was sometimes occupied by flight engineer René Lambert, and the aircraft was used to check performance with the front engine shut down and its propeller feathered. Receveau considered its behaviour in this configuration excellent, summing up the tests as follows:

'The purpose of the trial was to make a series of dives with the front engine stopped and the rear at idle. I then levelled the aircraft when it reached its maximum permissible speed, maintaining level flight while speed was progressively lost until the angle of attack led to the stall. Recovery and restarting the engines was not difficult.'

The two-seater was used to test manoeuvrability with both engines running and at high g-loading. Few problems were encountered, apart from some tail buffeting at high speed.

Do.335 number 1 continued to be used for tests of control efficiency, but a flight on 13 January 1948 was terminated because of erratic behavior of the front

Another view Do.335 'No.1' at Brétigny in 1947. By this time the cannon had been re-fitted.

propeller, which would not go into coarse pitch. On 5 March the general test programme was officially ended, and M14 made its last flight on 4 June. It was concluded that handling and longitudinal stability were good while flying on both engines or with either engine cut. Handling with only the rear engine running was slightly better, especially with regard to torque effect and below 500km per hour.

There is some confusion as to when M17 last flew – it was either on 25 November or 3 December 1948. The exact date remains a mystery because of inconsistencies in two reports. November 25 is mentioned in an incident report, but the same incident is recorded in the CEV's photographic archives as having occurred on 3 December. Whatever the date, the last flight saw Receveau, with Lambert in the rear seat, take off from Bretigny airfield at 1500 hours for a flight, lasting 65 minutes, which was uneventful until the aircraft was being prepared for landing. The pilot selected undercarriage down and noticed that the starboard leg indicator light showed unsafe, although the mechanical indicator on top of that wing disagreed. Receveau recycled the undercarriage several times, but with similar results. Reasoning that the indicator must be malfunctioning, he decided to land. The landing was normal and, having brought the aircraft to a stop at the end of the runway, he started to backtrack. After a mile the starboard main leg suddenly collapsed, damaging the wingtip, destroying the lower fin and bending the rear propeller blades.

The aircraft was examined to determine the cause of the failure. It appeared that a foreign body might have jammed a valve in the hydraulic system, preventing the leg from locking down, but, as the exact cause of the incident could not be established, the groundcrew was immediately exonerated. The last of the French Do.335s was not repaired.

Almost as a footnote to the testing, on 15 December a test was carried out on the explosive removal of the rear propeller and the upper fin, but neither worked. The reason was not discovered until some years later; a relay on the undercarriage prevented activation of the system when the legs were compressed! Both aircraft remained in store at Bretigny and were eventually scrapped, probably in March 1949.

The pilot's cockpit of No.2. The Revi EZ 42 gunsight had been replaced by a French SFOM version, as had some of the flight instrumentsation. [CEV via Hugh Jampton Collection]

Two views of Do.335 No.2 after the undercarriage retraction incident at the end of 1948. The rear prop and fin was damaged beyond repair, and the lack of spares were such that it was impossible to to put right. *[both CEV via Hugh Jampton Collection]*

Testing the Salamander

The Do.335 was not the only type examined, tested and flown by the French. The earliest document located in the French military archives on the Heinkel He.162 Salamander is dated 25 April 1945 and is report No 219, written by Capt Mirles, head of the *Mission d'Information Scientifique et Technique* (MIST), and his assistant, Capt Gentil. The report includes the statement that *'A member of the Luftwajfen Beute, captured near Gotha, carried documents giving detailed information on the construction of the Volksjager ('people's fighter') 162. Another PoW, who worked under command of the Chef Technische Luft Rustung from August 1944 to February 1945, had documents dated March 12, 1945, on the waterproofing and packing of the He.162's wooden components'.*

Although no He.162s were discovered in the French occupation zone, the British Forces of Occupation in Germany - which was to become the British Army on the Rhine (BAOR) on 25 August 1945 - gathered thirty one of these machines at Leck airfield in Schleswig-Holstein. According to the British Air Forces of Occupation report, ten were destroyed where they were found, twelve were kept, two given to the Americans and seven handed over to the French. Eventually, four He.162s were sent to the USA and five to France. Unfortunately, no record has been located to provide the date when the other two French examples were handed over to the Americans.

The five Salamanders destined for the French were loaded on goods wagons at Leck railway station by men of 83 Group, attached to 8302 Air Disarmament

Two of the captured He.162s seen partially dismantled and parked on a railway waggon at Leck, Germany in February 1946. Closest to the camera is 'Yellow 5', Werke Nummer 310003, while behind is Werke Nummer 310005 'Red 7'.

Wing of the RAF, which organised the move on behalf of a group of five officers, Cmdts Bouvarre and Hirshauer, Capts Petit and Boitelet and mechanical officer Sous-Lt Schall.

This group – which also included forty pilots of the Armée de l'Air - had the task of choosing and transferring to France a large number of miscellaneous German aircraft, including the He.162s. It was impossible to cross the Rhine by train before February 1946 because the bridges had been destroyed, so the five He.162s were stored in the Armée de l'Air Depot in Neuwied railway station, near Koblenz. On 2 March 1946, they arrived at Andernach railway station, on the French bank of the Rhine, and then left for the Établissement Central du Materiel Aéronautique de Nanterre (formerly Beute Park 5 of the Luftwaffe), which specialised in the recovery of airframes.

At the end of March the He.162s were transferred for study to the Société Nationale de Construction Aéronautique du Centre (SNCAC), in the former Farman factory at Boulogne-Billancourt.

Three He.162A-2 airframes were initially selected for restoration to flying condition, the other two He.162A-ls being kept for technical study.

The wooden wings of two He.162A-ls, fitted with 30mm MK 108 cannon, were dissected to analyse their wooden structure. These were Werke Nummer 310005 'Red 7' and Werke Nummer 310003 'Yellow 5', built by the Junkers site in Bernburg. Both had been flight-tested between the end of March and mid-April 1945 by factory test pilot Heinrich Osterwald, then transferred to Ludwigslust on 11 April by the pilots of I./JG1, and found in Leck by British troops. The two Heinkel ejection-seats were sent to Société Nationale de Constructions Aéronautiques du Sud-Ouest (SNCASO) for installation in the first French jet aircraft, the SO.6000-03 and -05 Triton, and a French-modified

'Yellow 1', Werke Nummer 120223 formerly of 3./JG1 and later to become the French 01 is seen at Ludwigslust in April 1945.

Heinkel He.162 - thought to be 'White 21' Werke Nummer 120015 of Stab1/JG1, and about to become the French No.2, seen at Ludwigslust during April 1945. The narrow track undercarriage is noteworthy.

version of the seat was installed in the SO.6020-01 and -02 Espadons. The four 30mm MK 108 cannon were passed on to the Direction des Etudes et Fabrications d'Armement (DEFA) for examination, and what remained of the A-1 airframes were held to provide spares for the flight-test aircraft.

The three He.162A-2s, armed with 20mm MG 151/20 cannon, were transferred by ground vehicle to Toussus-le-Noble airfield.These had been built by the Heinkel factory in Marienehe, were Werke Nummer 1200015 'White 21' or 'Yellow 21', Werke Nummer 120093 'White 2' – which had been ferried from Parchim to Ludwigslust by Oblt Hans Berger on 6 April 1945 - and Werke Nummer 120223 'Yellow l' - ferried on 15/16 April 1945, from Ludwigslust to Leck via Husum by Oblt Gerhart Stiemer. Oberleutnant Wolfgang Wollenwebner flew Werke Nummer 120223 at Leck on 22 April 1945.

Owing to the lack of maintenance and flight manuals, and also because the technicians and workers of SNCAC had to learn the new techniques peculiar to jet aircraft, restoration of the three He.162s to flying condition took a year. Meanwhile the Compagnie Electro Mécanique (CEM) in Paris was charged with overhauling six BMW 003 jet engines.

The Centre d'Expériences Aériennes Militaires (CEAM), based in Mont-de-Marsan, was created in August 1945, with Colonel Kostia Rozanoff its director. At that time the Ministére de l'Air and the Etat-Major General Air (EMGA) thought that the French aeronautical industry should be prepared to re-equip the Armée de l'Air with modern aircraft within the next five years. Rozanoff's role was to prepare the equipment for the operational units and tell them how to use it. Pending the arrival of French and British jets, Rozanoff wanted his pilots to familiarise themselves with the German jets that France had just received from its allies.

Following the arrival of jet engines in the Nanterre depot, a few Jumo 004s were sent to the school of mechanics in Rochefort to enable the mechanics posted to the CEAM to be trained in procedures. From August 1946 the *Bureau Etudes et Plans* (BEP) of the EMGA wanted the CEAM to have two jet aircraft at its disposal as soon as possible, like the two Me.262s restored to flying condition by SNCASO.

In September 1946, at the request of CEAM, director Lt-Col Badré, head of the BEP, sent four mechanics and a pilot on a training course on the BMW 003E engine at the CEM in Paris. Likewise, in October 1946, two aircraft mechanics and two engine mechanics of the CEAM were sent to Bretigny to witness the reassembly of two Me.262s by SNCASO, to familiarize themselves with German techniques before the Heinkels arrived.

The new 'first flights' of the Heinkels took place in April and May 1947, and were performed by factory pilots Abel Nicolle and Louis Bertrand at Orléans-Bricy airfield, which had a 7,050ft concrete runway. During these flights the three

He.162s received their French identifications : '01'; formerly Werke Nummer 120223 'Yellow 1' of 3./JG1; '2'; formerly WNr 120015 'White 21' or "Yellow 21' of 1./JG1; '3': formerly WNr 120093 "White 2" of 1./JG1.

At the STAé's request, following his first trials on 01, Abel Nicolle wrote a report for the CEAM on the procedure and handling of the He.162:

A - Before start:
Having had the correct functioning of the engine checked by the qualified groundcrew (check for leaks on the fuel lines etc), sit in the cockpit and check that;
 1) The flaps are set to take-off position (30°, seven/eight actions on the pump), then lock them;
 2) Elevator trim is set at about 1.5°; then
 3) Start the engine;
    a - start the Riedel [starter motor] and let it rev up to 800 r.p.m;
    b - open the fuel cock;
    c - press the ignition button on the throttle;
    d - When 1,200 r.p.m. is reached, push the throttle to 'ground idle' stop;
    e - Check the tachometer; if the revs grow steadily, stop the ignition on the throttle and check the temperature (max 750°C).

When 2,200 r.p.m. is reached, stop the Riedel, then the engine must settle on its ground-idling setting, i.e 3,800 r.p.m. Then take off after a short trial ground-run.

The take-off run is rather long, about 2,000m for an indicated airspeed (IAS) of 190-200km/hr; the aircraft has no tendency to swing. Retract the undercarriage as soon as possible, keeping the lever in the low position until the three wheels are locked (apply brakes to avoid vibrations). Do not retract the flaps before the safety altitude is reached; the aircraft has a strong tendency to sink during this operation. Do not exceed 350km/hr with flaps and undercarriage down.

B - Before landing
Reduce the speed to about 300km/hr, then set the flaps to the down position (start position -30°), and then lower the undercarriage. The engine speed being at about 6,000 r.p.m., the descent is rather slow. As soon as the pilot judges he is in the best situation for landing he should reduce the power below 6,000 r.p.m., slow the aircraft down to about 250km/hr, and set the flaps to their maximum 'down' setting (again this must be done at a safe altitude, as the aircraft sinks rapidly in these conditions). If necessary, add some power carefully to avoid any rough running. If the aircraft is kept straight and level the approach is made between 200-210km/hr. Landing is easy; keep the aircraft as long as possible on its main wheels and brake gently. If the landing run is too long, shut the fuel feed and set the throttle to 'idle'.

C - Important remarks regarding safety
Do not fly longer than 30min (including take-off and landing). In case of engine failure, the pilot should land with the undercarriage up if he is not absolutely sure he can reach the runway. If the engine stops it is impossible to retract the undercarriage. Therefore the pilot should not lower the undercarriage unless he is absolutely certain to touch down on the runway.

D - Performance
SNCAC, as requested, did not undertake performance measurements. It seems, however, that at an altitude of 4,000m and at an engine speed of 9,000 r.p.m., the airspeed should have been around 550km/hr. At the same altitude it should be around 610km/hr at 6,300 r.p.m. Time to climb to 4,000m at 9,500 r.p.m. and 320km/hr IAS was 7 minutes.

E - Fuel consumption
Climb to 4,000m at 9,500 r.p.m. and descent at 6,000 r.p.m.; 285lit; in level flight at 4,000m and 9,000 r.p.m.: 950lit/hr; Usable fuel capacity: 640 lit in aircraft No 01 (470 + 170 for reserve)".

On 8 May chief engineer Pierret of the STAé wrote to EMGA:
'Aircraft loaded as follows; take-off weight 2,713kg, c.g. at 22-6 per cent.
A - To begin with, I draw your attention to the undercarriage-up selection manoeuvre; after the port undercarriage is locked, the lever returns to its initial position, the left leg staying in the down position. It is therefore necessary to reactivate the undercarriage lever to retract the undercarriage fully.
B - Otherwise, the performance charts attached to the German manual seem to be very optimistic, especially for the take-off run, which in reality needs 1,200m with no wind; the landing takes 1,400m with flaps set at 25°. The take-off speed is 200km/hr IAS with the engine running at 9,500 r.p.m.
The aircraft does not seem very steady on its course. Especially when landing with under- carriage and flaps down, the aircraft is difficult because of its lack of stability in the yaw and pitch axes.
C - Considering the shape of the fuel tanks, it is advisable to keep a safety margin of 200lit of fuel for descent and landing.
Under these conditions it is compulsory to limit the duration of a level flight at 4,000m (i.e. between 250 and 350km of flying range with no wind) to l5min. It would therefore be wise to avoid intermediate stops when ferrying the aircraft to Mont- de- Marsan."

Following the first factory flights performed by SNCAC, a team from the Direction Technique et Industrielle (DTI) was tasked with testing this aircraft at the request of the EMGA. This team comprised chief engineer du Merle, Capt Descaves commanding the Reception group in the Centre Aérien Technique de Reception et d'Entrafnement (CATRE), Capt Le Martelot and Lt Lombaert.

Lt Raphael Lombaert, a former Spitfire pilot with 329 Sqn on D-Day, and used to German instruments owing to his delivery flights of the Focke-Wulf Fw.190 for SNCAC at Auxerre-Moneteau airfield on 18/19November 1945, found himself in a somewhat familiar cockpit. He reported;

"On 27 May 1947, after the display flight by Louis Bertrand of SNCAC, it was my turn to fly Heinkel He.162 No 2. Regarding the airframe, the total absence of documentation and technical data, apart from Abel Nicolle's report, did not make ground training easier.
I was used to the instruments, except some novelties such as the thrust-nozzle temperature gauge, known as T4. As for the levers, besides the usual throttle there was a lever for opening the fuel circuit for the engine, and, very close, the undercarriage retraction lever; only retraction, as lowering was made thanks to a corkscrew-type handle linked to a cable. The action on this handle unlocked the locking mechanisms for the legs, which, upon retraction, had tightened big springs, the purpose of which was to pull the three wheels back to the down position. In a way it was an emergency lowering system without hydraulic pressure. Lastly, near the left-hand side of the windscreen, was the hand pump which allowed the flaps to be lowered hydraulically for take-off and landing. Turning this handle in one direction allowed them to be locked into position and, in the other direction, to be unlocked for retraction.

One of the French He.162s awaits flight.

One of the He.162s, thought to be No.2, in full French colours and markings.

Lastly, for these flights we did not have any oxygen or radio. A very competent Austrian motor engineer briefed us very precisely on the handling of the BMW 003 turbojet, its starting with the help of the Riedel motor, its reference settings, its temperature limits and, above all, issued the utmost precautions to use the throttle towards the maximum thrust without pumping or flaming out. It took about 20-25sec to go from idle to maximum power. In fact, three characteristic settings were used: Ground idling at 3,000 rpm when declutching the Riedel; flight idling at 6,000 rpm (T4 of 450°C); and maximum power at 9,800 rpm (T4 of 600°C, which should not exceed 700°C). Of course, any intermediate setting was possible between 9,800 and 6,000 rpm. The BMW 003 was very sensitive to negative g and flamed-out easily.

"I shall not dwell on the take-off, flaps on the proper setting thanks to seven actions on the pump, but I stress the sluggish acceleration, which requires a run of 900-1,000m before the aircraft leaves the ground at a speed of 190km/hr. The available power was the equivalent of 650 hp maximum for a take-off weight of 2,800kg and a wing loading of 230kg/m2. Climb at 350km/hr was very pleasant, with no vibration and a very soft noise; levelling off the aircraft at the cruise setting of 9,000 rpm and 480km/hr IAS did not alter the pleasure of the climb. Unfortunately this apparent docility vanished as soon as one started to feel for the control reactions, especially its incredible sensitivity to the slightest amount of slip, which made one think, because of the sinking in pitch or roll, that a spin was imminent.

In addition, a slightly tight turn caused a worrying buffeting. A speed run 'all in the middle' at 9,600 rpm brought the IAS in the vicinity of 750 km/hr at 1,500m. In all, this aircraft was very pleasant in 'recreational' use, but could become terribly vicious in manoeuvres relatively common for a classic fighter.

The limited 15min of flight imposed as a precautionary measure, not taking account of the real fuel consumption, had almost elapsed, and it was about time to play Act 5, the landing. Because of the engine characteristics this was one of the most difficult phases of the flight and, in my opinion, must have been the cause of several accidents.

1st manoeuvre; Slightly reduce the normal flight setting to flight idling, the speed decreasing to 260km/hr.

2nd manoeuvre: Lower the undercarriage and pump on the flaps; pump 21 times, the speed settling at 240km/hr at 6,000 rpm. (flight idling).

3rd manoeuvre; make the last turn and estimate (in view of the runway) the right moment to set the throttle to ground idling. From then on it is not possible to add more power to lengthen the landing or make a go-around, because in the following 20sec (necessary to regain full power from the ground-idling setting), the aircraft would have travelled, at decreasing speed, between 1,300m and 1,100m.

If everything went right, flying past the approach lights at the runway threshold at approximately 200km/hr led to touchdown within the following 200m. The aircraft had to be maintained 'nose-up' on its main undercarriage and, bearing in mind the roughness of the brakes, the nosewheel had to be put down at 100-110km/hr before applying the brakes. Then, with the help of the residual thrust, I returned to the end of the runway, where a van was waiting to refuel the aircraft using the ancestral method of the Japy hand-pump plunged in a 200 litre barrel, pouring the amount of precious fluid necessary for the following flights into a funnel capped with a chamois leather."

Lombaert found the visibility from the cockpit very poor. His report went on to say 'The pilot can't see in the crucial area – that is to the rear and to the top. Moreover, the very short range and the scanty equipment makes flights in bad weather or flying through the clouds somewhat problematic'.

Two incidents disrupted the following flights by Le Martelot and Descaves of CATRE. During the take-off run the aircraft, having reached the normal take-off point, kept on rolling, leaving the ground just short of the end of the runway and frightening the spectators. The pilot, noticing that the aircraft refused to take off, realised that the flaps had not remained in the take-off position because he had forgotten to lock their handle by a quarter of a turn before applying the take-off power, which he rapidly corrected. The next incident led to subsequent flights being cancelled, as, during the sixth go-around on landing, an undercarriage door was found partly torn away. On debriefing, the pilots came to the conclusion that the aircraft was potentially dangerous because of its unexpected reactions, and tricky on landing.

In October 1946 Colonel Kostia Rozanoff left CEAM to take charge of flight testing for Marcel Dassault. Lieutenant-Colonel Longuet succeeded him at Mont-de-Marsan, his deputy director being Commandant Philippe Maurin, also responsible for the Fighter section. The head of the design department was Cmdt Emile Thierry, also responsible for the CEAM's Bomber section.

On 17 May 1947, following Abel Nicolle's test-flying report, Col Georges Grimal of the Bureau d'Applications Téchniques (BAT) transferred the three He.162s by rail, and asked the director of CEAM to send the four mechanics attending the SNCAC training course in Orleans-Bricy to engineer M Siriex's place, and to have a pilot make contact with personnel from SNCAC who had flown this aircraft type, in order to gather all the necessary information before the tests.

Two months later, after the go-ahead had been received from the EMGA, the three Heinkels were reassembled by the base's repair and overhaul workshop, with the help of the civilian personnel of SNCAC, a former engineer from BMW and a new team of CEAM mechanics specialising in jet aircraft. The He.162s were ready to fly in September 1947. On 25 October eleven pilots flew about 20 minutes each to familiarise themselves with high speeds and the use of jet engines.

One problem was the starting of the Riedel two-stroke starter engine by pulling a cord. Attempts sometimes ended with the aircraft back in the hangar, the Riedel having refused to start. The clearance between the turbine blades and

Not the best of images, but one that shows He.162 French No.3 undergoing engine runs at Salon-de-Provence. The aircraft had been repainted in French 'roundel-blue' overall with a 'wine' coloured nosecone.

the exhaust nozzle ring had to be checked regularly. If a blade was found to have lengthened, this had to be offset by grinding the ring.

The value of the tests was thought to be limited due to the lack of the oxygen feed and radio, and the aircraft's very short range and unknown reliability. The He.162, had been constructed from poor quality non-strategic materials, and suffered from other shortcomings; one was that the steel engine fasteners could not withstand the heat from the nozzle, and tended to sag under the stress of landing.

Flight familiarization took several days, proceeding by trial and error, because there were no technical manuals. Recorded pilots impressions of the cockpit were not the best; a cramped cockpit, containing the bare minimum, like the basic instruments. The throttle levers were simple iron rods, and strings with pieces of wood tied on their ends were the handles for the flaps and the undercarriage. The ejection seats had no cartridges, so did not work. Clearly the aircraft was a complete novelty for the pilots, none of them having flown a jet before. Once the starting difficulties were resolved, both the temperamental Riedel and the engine itself, they regarded the He.162 as being a very light machine, that was very easy to taxy with the brakes and not vicious on take-off, with the exception of a high sensitivity of the ailerons and the narrow track, which made it react to the action of the control column before take-off speed was reached. In flight the aircraft was smooth, but as far as is known, no one attempted aerobatics. The biggest concern was always the very short flight time, which prevented anyone from climbing to altitude or moving away from the airfield.

Heinkel No 2 was slightly modified in 1948. A second pitot tube was added above the first one, as well as a small aerial in a fairing on the starboard side of the nose. It also lost its weapons and the radio antennae on top of the engine.

On the morning of 23 July 1948, Capt Schlienger, who had already logged 25min on the type, prepared for his second flight, to display the Heinkel 01 to the officers from the superior headquarters course in Mont-de-Marsan. At about 1115hr the aircraft was towed to the end of runway 10. Schlienger sat in the cockpit, and the flaps were lowered to the take-off position. At 1125hr the pilot started the engine and after going from ground idle to flight idle the chocks were removed and Schlienger released the brakes and taxied on to the runway.

All who witnessed the flight thought that the aircraft was not accelerating fast enough after having covered nearly 1,500m. The witnesses on the ground then lost sight of it, but the runway officer, Lt Dussaule in the control tower, also felt that the aircraft was slower than usual. At the end of the runway Schlienger laboriously took off, gaining just a few feet without retracting the undercarriage. The pilot attempted to land in the middle of a clearing, turning to port with undercarriage still down, then stalled and crashed into the ground, the aircraft immediately bursting into flames. The aircraft was destroyed, its engine being thrown more than 60m from the point of impact. No mechanical failures were found; the failure was probably electrical.

Heinkel He.162 No 0l had logged 7hr 45min since its overhaul in France, and its engine had logged a total running time of 7 hours 55 minutes since its last overhaul by CEAM. Following the publication of the accident report on 30 July 1948 it was concluded that these aircraft were no longer of any military value and so the other two machines were written off.

Chapter Eleven
Japan

Virtually as soon as the dust from the two atomic bombs had settled, the Americans sent over teams of investigators to plunder Japanese technology. Not only did they sift through the wreckage, they also interrogated surviving designers, engineers and Air Force personnel. From virtually the moment American feet touched Japanese soil a steady stream of information, parts and aircraft began arriving in the USA.

The main batches of Japanese aircraft selected for evaluation in the USA after the surrender were concentrated at Yokosuka Naval Base, near Yokohama. If possible, the aircraft were made serviceable by Japanese personnel, before being ferried by Japanese pilots to Oppama airfield at Yokosuka under escort by US military aircraft. In a few cases incomplete aircraft were finished under US supervision before flying to Oppama. Some aircraft, such as the suicide types, were crated where they were found and shipped as cargo in the hangars of aircraft-carriers. Most aircraft made the journey on aircraft-carrier decks, since they were too large to be taken below deck on the small lifts of the vessels concerned.

The shipments to the US were on board escort carriers USS *Core* (CVE-13) and USS *Barnes* (CVE-20) during November, and the USS *Bogue* (CVE-9) during December 1945. The plan was to obtain four examples of each type which had not previously been evaluated during the war. One example was to be for the USAAF, one for the USN, one for the RAF and one for later allocation or for use as a source of spares.

The Japanese delegation at the formal surrender on 2 September 1945 when representatives from the Empire of Japan signed the Japanese Instrument of Surrender in Tokyo Bay aboard the USS *Missouri*. (USAAF)

The three escort carriers each carried a large number of captured aircraft. In addition the large H8K2 Emily flying-boat was placed on board seaplane carrier USS *Cumberland Sound* (AV-17), also sailing from Yokosuka. The *Barnes* arrived at Norfolk, Virginia, on 7 December 1945 and commenced unloading forty-five Japanese aircraft on 10 December for transfer to the Naval Air Station. The *Core* left Yokosuka on 16 November 1945 and docked at Alameda before returning to the Pacific.

It was intended that the USAAF aircraft would be delivered to Langley Field, Virginia, to be made airworthy for transfer to Freeman Field. Personnel from Freeman Field had already arrived at Langley Field to work on the first batch of aircraft when their orders were rescinded. Due to a change of policy by Air Materiel Command order TSMCOZA-2-109 of 13 February 1946, it was stated that the Middletown Air Materiel Area (MAMA), Olmsted Field, Middletown, Pa, would become the sole Air Materiel Command repair agency for Foreign Equipment, including items designated as 'Class 32' for museum use.

The first Japanese aircraft arrived at Middletown from Langley Field on 19 January 1946 and, by 30 June, fifty complete aircraft, and parts of nine others, had been received there from Langley Field and Newark. In addition, three aircraft had been flown out from MAMA to Wright or Freeman Fields and two others had been flown out of Newark after overhaul by MAMA personnel sent there on temporary duty.

Investigating developments.

Unlike the Germans, who for many decades had a reputation of being in the forefront of technology, it was not until the 1941-45 fighting that the Japanese gradually lost their reputation for being a nation of copyists. This was especially

It was not just Germany that was littered with abandoned aircraft - the American investigation teams had plenty of wreckage to choose from in Japan as well! *(USAAF)*

A partially wrecked Nakajima G5N Shinzan 'Mountain Recess'. This was a four-engined long-range heavy bomber designed and built for the Imperial Japanese Navy, with a designation of 'Experimental 13-Shi Attack Bomber'; the Allied code name being 'Liz'. It was photographed at Atsugi airfield, Japan on 5 September 1945. *(USAAF)*

All Aboard!
Above: The fourth of four prototypes of the Nakajima G8N1 'Rita' four-engined heavy bomber. This example was incomplete at the end of the war at the Nakajima factory at Koizumi, north-west of Tokyo. The aircraft was completed and flown to Yokosuka on 7 December 1945, to be shipped to the USA aboard the escort carrier USS *Bogue*. The 'Rita' was off-loaded at New York during January 1946 and taken by barge to the AOAMC at Newark AAF.

Below: This Kawanishi H8K2 Model 12 'Emily' Navy Type 2 Flying-Boat previously belonged to the 801st Naval Air Corps based at Takuma, Shikoku. It was flown from Takuma to Yokosuka on 13 November 1945 to be shipped to the USA on board the seaplane tender USS *Cumberland Sound*, being off-loaded at NAS Whidbey Island, Washington State. Somehow - the means has not been recorded - it crossed the continent to be tested by the US Navy at NAS Norfolk and NAS Patuxant River. It made only one recorded flight from NAS Norfolk to ATC Patuxent River on 23 May 1946, afterwards being relegated to taxiing trials because of engine problems. These trials, under Project No.PTR1411, were aimed at studying the type's hydrodynamic stability and characteristics of the spray generated during take-off. The tests took place between 22 August 946 and 30 January 1947, eventually being terminated by failure of an engine which could not be replaced. It was later put into storage at NAS Norfolk on behalf of the Smithsonian Institution. It was returned to Japan in July 1979 and is now on display in the Museum of Maritime Science at Tokyo.

Major Carroll R West examines a The Kyushu Q1W Tokai 'Eastern Sea'. This was a land-based anti-submarine patrol bomber aircraft developed for the Imperial Japanese Navy. The Allied reporting name was Lorna. Although similar in appearance to the German Junkers Ju.88 medium bomber, the Q1W was a much smaller aircraft with significantly different design details. According to the original caption, this picture was taken on Mizutani Airfield, Hokkaido, Japan.

true in the field of aircraft design, because high-performance models of home construction quickly scored major successes and dispelled the view that Japan had produced only outmoded copies of foreign types. The Americans were especially interested in discovering how much was home-grown, and how much was not.

A number of the Japanese aircraft appeared to have been derived from Messerschmitt machines - though they were not exact copies. It was also known that further original work had been carried out by the Japanese on power plants.

The Navy investigates

The U.S. Naval Technical Mission to Japan (NAVTECHJAP) was established on 14 August 1945 by the Chief of Naval Operations, in accordance with the Intelligence Appendix of Operation Blacklist, the operational plan for the occupation of Japan. Capt. Clifton C. Grimes, Fleet Intelligence Officer in Charge of Technical Intelligence for Joint Intelligence Center, Pacific Ocean Areas (JICPOA), was designated chief of mission. The nucleus of personnel came from among those attached to JICPOA who had technical and language qualifications and from technical personnel at other commands. The initial group was designated JICPOA Team No. 29 and entered Sasebo harbor on 23 September on board the attack transport *Shelby* (APA 105) on the date of the initial occupation of Kyushu, Another group, designated JICPOA Team No. 30, joined the Third Amphibious Group in the occupation of certain areas of China. Elements of the intelligence groups of Commander Seventh Fleet joined in Sasebo, and on 28 September all units were consolidated as NAVTECHJAP. The headquarters, initially located at Sasebo, was soon moved to Tokyo to improve coordination with the other occupation activities.

The purpose of the mission was to survey all Japanese scientific and technological developments of interest to the Navy and Marine Corps in Japan, China, and in Korea south of 38° north latitude. The mission's work involved seizure, examination, and study of intelligence material, interrogation of personnel and preparation of reports.

Before the cessation of hostilities, ONI had prepared a list of all the Japanese

It was not just rockets and jets the Americans were interested in - this Nakajima B6N Tenzan 'Heavenly Mountain' (Allied reporting name: 'Jill') was the Imperial Japanese Navy's standard carrier-borne torpedo bomber during the final years of the war and the successor to the B5N 'Kate'. It was fitted with radar, the aerials being on the wing leading edge and along the rear fuselage. At least one B6N2 was taken to the USA and tested at NAS Anacostia by US Navy personnel of the Technical Air Intelligence Center there. (both USAAF)

technical 'targets' it desired, including lists of specific items and information sought by the technical bureaus of the Navy Department. As early as 15 September, copies of 'Intelligence Targets Japan' of 4 September 1945, prepared by ONI, were received by NAVTECI-LIAR permitting the movement of the mission without much additional planning.

NAVTECHJAP was organized into two departments: Executive (administration, etc.) and Technical. The latter was divided into sections; Ships, Electronics, Ordnance, Medical, Special, and Petroleum. One other section had the job of filing, printing, editing, and distributing intelligence material.

The Technical Liaison Section, located at the Intelligence Staff (G-2), Supreme Commander Allied Powers (SCAP), attended policy conferences and other meetings and maintained contact with SCAP headquarters. The Special Intelligence Section exploited any non-technical targets that might be assigned. It also assisted in the completion of the U.S. Strategic Bombing Survey after the departure of the survey's personnel from Japan.

By 1 November 1945, NAVTECHJAP had 295 officers, 125 enlisted personnel, and 10 civilian technicians assigned to it. Among the officers were approximately twenty-three British technical specialists and language officers.

Collection centres were established at Sasebo, Yokosuka, Kure, and Kobe for documents and equipment. Field personnel wrote reports, and the NAVTECHJAP headquarters in Tokyo edited, typed, and/or printed the reports after checking them for completeness, accuracy and acceptability.

Japanese jets
The Japanese study of jet engines began in about 1930. One of the leading engineers in the subject was a navy officer, Dr. Tokutasu Tanegashima. Mitsubishi made a few small impulse duct units before the war but without remarkable success although he had been supported by Admiral Matsukasa in his endeavours.

A poor quality, but rare picture of the TSU-11 compressor engine in a test stand.

It has been claimed that Tanegashima visited Europe and the USA in 1940 as he had been motivated by information being made available on turbine technology. As a result, in 1941 the Institute of Naval Aeronautical Technology established a Thrust Section in its Power Plant Division at Yokosuka, creating a power unit that followed the Campini principal. This consisted of a a 130 hp air-cooled Hitachi Hatsukaze HA-11 ('Fresh Wind') four cylinder piston engine driving a geared-up ducted fan as a compressor, after which extra fuel was burned in an exhaust duct. Vanes straightened out of the airflow from the compressor before the extra fuel was injected.

The TSU-11 is said to have had a static thrust of 441 lb at 3000 rpm (the fan itself rotating at 9000 rpm), for a weight of about 200 kg. The TSU-11 was used in the Oka Model 22, and one was flown on a 'Ginga' Land-based Attack Aeroplane (stated in one source to be a Model 13) as a jet booster for assisted take off and emergency high-speed operation.

Nevertheless, other, better designs were just available at the end of the war. Another early jet unit was the TR-10, sponsored by the Navy, but bench tests were not very successful. The first Japanese turobojet series were designated TR for 'Turbine Rocket'. Early in 1945 TR designations were changed to NE for 'Nensho' or combustion rocket. It was developed around a centrifugal compressor that was adapted from a turbocharger, and a single-stage turbine. A bench test of the TR-10 was made in 1943 but the compressor gave a pressure ratio of only 3.5:1, when 4.1 had been expected and overall efficiency of the engine was only about 50%.

A new design incorporated a four-stage axial compressor and a new inlet duct in front of the centrifugal compressor, with a view to easing the load on it. The turbojet was designated TR-12. When completed at 772 lb, it was found to

Another poor quality but important image is this picture of a Kawasaki Type 99 (Ki-48) 'Lily' light bomber with what is said to be a Ne-00 gas-turbine engine slung underneath.

245

be heavier than the thrust it produced. It was then refined to lighten it and was designated the TR-12b.

Japanese records suggest that the TR-12b fuel consumption was 1,125 lb per hour and that it was planned to flight test it beneath a Mitsubishi G4M2 Betty bomber in November 1944. Confusion exists as to if this was ever done, but photographs are available that show a Kawasaki Type 99 (Ki-48) light bomber with what is said to be a Ne-00 gas-turbine engine slung underneath.

The Ishikawajima Heavy Industry Corporation carried out research on turboprop units, but official interest was slight. At the end of the war design of a 5,000-h.p. turbo-prop powerplant had started.

Small solid and liquid rockets were produced for the Oka and for various missiles, and in 1944 the Japanese acquired the manufacturing rights of the Walter 109-509 rocket unit, plus one example and the drawings. For this they paid 20 million Reichsmarks. Mitsubishi, and later, the Yokosuka Air Technical Depot, developed this unit, the Japanese version being known as the Toku Mark 2 and also as the KR 10. It gave about 3,307 lb. of thrust. Turbojets of the 'Ne' series were the main hope of the Japanese forces, and they were in the Special Attack Aeroplane 'Kikka' and intended for late Oka variants.

In the autumn of 1943, the first domestic gas-turbine engine, the Ne-00, was ready to be tested. Its flight trials were carried out under the fuselage of a Type 99 Twin-motor Light Bomber (Ki-48), but it and the Ne-10 were dropped for the Ne-12, an axial-flow unit giving about 705 lb. of thrust, which was completed in autumn 1944.

However, trouble was experienced with this unit as the turbine blades were prone to crack and the burner had to be enlarged and the turbine shaft lengthened. The necessary research threatened to delay successful development, and a new unit, the Ne-20, was hurriedly prepared. It was developed by co-operation between the Navy, Nakajima, Mitsubishi, and the Ishikawajima Shibaura Turbine Co. It was actually a scaled-down version of the BMW 003, though the Japanese had only photographs to work on. These arrived in August 1944 in the last liaison submarine from Germany; detailed handbooks were in another submarine, but this was sunk by a mine near Singapore.

The Ne-20 was ground-tested in March 1945 and as considerable success was attained it was put into production and installed in Kikka prototypes. These units had eight-stage axial compressors. The Ne-20 went into small-scale

The Mitsubishi G4M (or 'Type 1 land-based attack aircraft') was the main twin-engine, land-based bomber used by the Imperial Japanese Navy Air Service. The Allies gave the G4M the reporting name 'Betty', while Japanese Navy pilots called 'Hamaki' (cigar), due to its cylindrical shape. This example appears to have been found abandoned and damaged. The G4M carried the Ohkas in a very similar manner to how the Americans used a B-29 to get the Bell X-1 experimental rocket aircraft airborne a few years later.

Original caption:' *Jap hangar storing 'Bukkabomb' with 5th Air Force officer standing alongside to show comparative size. JAPAN Sept. 1945'.* (USAAF)

production and the Ne-130 and Ne-230 were later projects which were not built.

Special Attack Aeroplane 'Oka' Model 11

The idea of large-scale suicide attacks as a normal method of warfare was born in Japan in 1944. Although ordinary operational types were used in many suicide attacks, plans were put in hand for the use of a specially-designed piloted missile with characteristics specially suited for this type of warfare. The small Special Attack Aeroplane 'Ohka' Model 11 - a rocket-propelled monoplane, designed by Captain Niki, was the result. 'Ohka' was reported in Allied intelligence summaries as a trainer for some time before its true nature was known. Intact specimens were recovered at Katena airlield on Okinawa.

The prototype Ohka flew in the autumn of 1944 and the glider trainer version (MXY7) flew in that year also. In some Japanese circles the aeroplane was called 'Jinrai'. It was not unlike a winged torpedo in appearance, and its small size made it a difficult target. The Model 11 was considered to be slow, and production was planned to go on to better and faster models.

In all, 755 of this model were built-155 at Yokosuka and 600 at one of the Naval Air Depots, between September 1944 and 11 March 1945, when production ceased.

The Ohka was split into a number of components for sub-contracting; the fuselages and war-heads were probably made at the Kanagawa Branch Depot

An Ohka is discovered in a wooded area of Japan. (USAAF)

of the Yokosuka Air Training Department where the explosives and rocket charges were installed. Wings and tail units were produced by Fuji Hikoki K.K., metal castings were made in Osaka and the Hiroshima area, and other components in Osaka and Tokyo. After trial assembly the aircraft were dismantled and sent to the front.

Some three hundred Ohka Model II aircraft and MXY7 gliders had been sent to Okinawajima, but by the time U.S. troops arrived, many had been destroyed in the bombardment. Some were in underground hangars and five were recovered in good condition. Some were a blue-grey colour overall and some were finished in green above and light grey below. A low, three-wheeled dolly was used for handling the Ohka on the ground, as it had no landing gear of its own. This dolly had an 18-inch lift, so as to raise it into the bomb-bay of the parent aeroplane. The only type which was encountered as a carrier for the Ohka was the Type 1 land-based Medium Model Attack Aeroplane (G4M2).

Technically, the Ohka ll was a low-wing single-seat rocket-propelled monoplane with twin fins and rudders. It had an high-explosive war-head containing 2,645 lb. of trinitroanisole, with an air-armed nose fuse and four base fuses. The pilot could not control the fuse safety wire when in flight.

Power was provided by three fuselage and occasionally two additional wing rockets. These were Type 4 Mark l Model 20 units, each with three cylindrical canisters of solid fuel, about 6 feet 6 in. in overall length and ignited electrically. The total thrust seems to have been about l,763 pounds for 8-10 seconds.

The Ohka was held in the bomb-bay of the G4M2 before air-launching. The pilot was provided with a speaking-tube and earpiece and a walk-around oxygen bottle for inter-communication. There was a system of warning lights operated from the mother plane. The instrument panel had a rocket switch with five positions-off, wings, and Nos. 1, 2 and 3 fuselage; an A.S.I. graduated from about 100 to 680 mph; an altimeter; a compass; and an inclinometer reading 5 degrees above and 25 degrees below the horizon. The pilot aimed his missile- and himself-with a ring and post sight just forward of the cockpit.

The Ohka was air-launched at about 174-202 mph and was capable of gliding some 52 miles at 230 mph from 27,067 feet. The wing rockets were often used for the long-distance approach towards the target.

A Yokosuka MXY7 Ohka K1 Trainer discovered at the Yokosuka Naval Air Arsenal 1945. *(USAAF)*

The cockpit of a Yokosuka MXY7 Ohka baka rocket bomb, photographed at Atsugi after the end of the war with Japan in 1945.
(USAAF)

Maximum range could be attained at a gliding angle of 5 degrees 35 minutes. The fuselage rockets were generally switched on for the final approach of about three miles, the level speed becoming about 534 mph. At an angle of 50 degrees or more the speed went up to 618 m.p.h., at which velocity good penetration was achieved and immense damage caused.

The assembly of the Ohka Model ll was very simple. It was made in six parts-fuselage, wings, tail unit, war-head, bomb fairing, and power unit. Wings were made of wood, and the fuselage of aluminium alloy with flush- and round-head riveting. The war-head was coupled on by four bolts; between its base and the propulsion unit there were eleven equally spaced circular formers. The pilot was provided with small pieces of 1-inch thick armour plate behind his head and under his feet.

It seems that the Japanese considered the Ohka Model 11 to be rather slow and vulnerable-it is true that its gliding approach was not very fast-and on 11 March 1945, production ceased in favour of later models, though these had not been entirely satisfactorily developed at the time of the surrender.

From this developed the 'Ohka' Model 22 which was an attempt to improve

A pair of Model 43 K-1 Kai rocket assist trainers, note the landing skid

the performance by utilising higher-powered thrust units was the development of Model 22, which had a TSA-11 gas-turbine motor.

The Ohka Model 22 was also considered as an interim type until something better came along, as its top speed of about 265 mph meant it was unable to avoid enemy fighters. It was apparently not used in action. The first one was carried to about 10,000 feet by a 'Ginga' but when dropped for flight trials it stalled and did not recover from the resulting spin.

The Yokosuka A.T.D. built fifty airframes but they were not entirely completed at the end of the war. Captured documentation suggested that Aichi had intended to make two hundred of this model, but were unable to do so because of war damage to the factory. According to an American report, a number of Ohka 22s were produced as two-seat trainers, without war-heads.

A close-up of the rocket tubes fitted to the Ohka.

An Ohka is examined by US troops. Clearly, although this device is similar to the German V-1, it really was a totally different design. *(USAAF)*

Accounts of the 30-Series 'Ohka' are somewhat sketchy. One report says that the Model 33 was to have been generally similar to the Model 22, and another source states that it was to have been an improved version with a Ne-20 gas turbine, to be launched from an Experimental Land-based Large Model Attack Aeroplane 'Renzan' (G8Nl). It was passed over in favour of research for the Ohka Model 43, which was an anti-invasion weapon with increased power. This was designed around the Ne-20 gas turbine unit. It was to have been built by Aichi, who carried out the stress analysis in May 1945.

The Americans discovered that it had been intended to use ground-catapult launching and skid landing, which suggested that the Japanese contemplated using it as a lightweight fighter with two cannon, rather than as a pure suicide aeroplane. At least one Ohka Model 43 was under construction, but was not completed by the end of the war. A trainer version was also under development for this version, the two-seat Model 43 K-1 Kai *Wakazakura* (Young Cherry), fitted with a single rocket motor. In place of the warhead, a second seat was installed for the student pilot. Two of this version were built.

Captured documentation revealed that there was one further Ohka project - the 50 series. It was to have been towed aloft by a tug aeroplane and released over the target. The project is believed to have been dropped as all Japanese runways were too short.

Experimental Design Local Fighter Aeroplane 'Shusui' (J8Ml) and Ki-200

The J8M1 was intended to be a licence-built copy of the Messerschmitt Me 163 Komet. Difficulties in shipping an example to Japan meant that the aircraft eventually had to be reverse-engineered from a flight operations manual and other limited documentation. A single prototype was tested before the end of the war.

The Japanese were well aware of the results of the strategic bombing of Germany, and knew that the B-29 Superfortress would be bombing Japan and the problems which would arise from trying to combat this. Japanese military attachés had become aware of the Komet during a visit to the Bad Zwischenahn airfield of Erprobungskommando 16, the Luftwaffe evaluation squadron charged with service testing of the revolutionary rocket-propelled interceptor. They negotiated the rights to licence-produce the aircraft and its Walter HWK 509A rocket engine. The engine license alone allegedly cost the Japanese 20 million Reichsmarks.

The agreement was for Germany to provide the following by spring 1944:
- Complete blueprints of the Me 163B Komet and the HWK 509A engine.
- One complete Komet; two sets of sub-assemblies and components.
- Three complete HWK 509A engines.
- Inform Japan of any improvements and developments of the Komet.
- Allow the Japanese to study the manufacturing processes for both the Komet and the engine.
- Allow the Japanese to study Luftwaffe operational procedures for the Komet.

The broken-down aircraft and engine were sent to Kobe, Japan in early 1944. It is probable that the airframe was on the Japanese submarine RO-501 (ex-U-1224), which left Kiel, Germany on 30 March 1944 and was sunk in the mid-Atlantic on 13 May 1944 by the hunter-killer group based on the escort carrier USS *Bogue*. Plans and engines were on the Japanese submarine I-29, which left Lorient, France on 16 April 1944 and arrived in Singapore on 14 July 1944, later sunk by the submarine USS *Sawfish* on 26 July 1944 near the Philippines after leaving Singapore.

The Japanese decided to attempt to copy the Me 163 using a basic instructional manual on the Komet in the hands of naval mission member Commander Eiichi Iwaya, who had travelled to Singapore in the I-29 and flown on to Japan when the submarine docked.

The Mitsubishi design staff began to translate the design requirements to Japanese production methods. In June 1945, two examples were ready for testing, one by the Navy as the J8M1 and another for the Army (Ki-200). A glider version (MXY8) was also made.

The J8M1 underwent ground testing during June 1945, and at Yokosuka on 7 July it made its first test flight, piloted by Lt.-Comdr. Inuzuka. After a steep climb to about 1,300 ft., however, the rocket unit ceased to function because air had entered a fuel pipe due to the steep angle of take-off.

The 'Shusu' fell almost vertically to the ground and was heavily damaged; Lt.-Comdr. Inuzuka died next day of his injuries.

The A.T.D. at Yokosuka began production of the Ki-200 but only about seven examples had been completed at the time of the Japanese surrender.

Data for the Shusui were: Single-seat interceptor fighter with jettisonable wheels, and landing skid, powered by one KR 10 (or Toku Mark 2) rocket unit derived from the HWK 109-509 and giving about 3,307 lb. of thrust. Armed with two 30 mm cannon in the wing roots.

The sole Mitsubishi J8M1 prior to flight.

As with many other Japanese fighter types combating air raids over Japan, the Ki-200 was considered for use in ramming B-29s. The envisioned mission profile was to make one or two firing passes and then, with the remaining energy, conduct a ramming attack. Any fuel left on board would most likely detonate, increasing the effectiveness of the attack, but also meaning the pilot had little chance of survival.

The Kyushu J7W1 Shinden
Another esoteric aircraft investigated by the Americans was the Kyushu J7W1 Shinden 'Magnificent Lightning' fighter which was built in a canard design. The wings were attached to the tail section and stabilizers were on the front. The propeller was also in the rear, in a pusher configuration.

Developed by the Imperial Japanese Navy (IJN) as a short-range, land-based interceptor, the J7W was a response to B-29 Superfortress raids on the Japanese home islands. For interception missions, the J7W was to be armed with four forward-firing 30 mm cannons in the nose.

The Shinden was expected to be a highly manoeuvrable interceptor, but only two prototypes were finished before the end of war. A gas turbine–powered version was considered, but never even reached the drawing board.

After the end of the war, one prototype was scrapped; the other J7W1 was claimed by a US Navy Technical Air Intelligence Unit in late 1945, dismantled and shipped to the United States. There appears to be some confusion as to which it was - some sources claim that the USN took the first built while others state that it was the second.

The sole remaining J7W1 was reassembled, but has never been flown in the United States; the USN transferred it to the Smithsonian Institution in 1960.

Finally, although many other aircraft were captured from Italians, Germans and Japanese, mention should be made of possibly the largest aircraft captured and evaluated.

In February 1943 the Imperial Navy staff asked Nakajima Aircraft Company to design a four-engined bomber, capable of meeting an earlier specification set for a long-range land-based attack plane. The final specification, issued on 14 September 1943, called for an aircraft with a maximum speed of 320 knots able to carry a 4,000 kg bomb-load 2,000 nautical miles or a reduced bomb-load 4,000 nautical miles.

Nakajima's design featured a mid-mounted wing of small area and high

The remains of the canard Yokosuka MXY6 research glider built to test the configuration of the Kyushu J7W Shinden, seen at the Atsugi Naval Air Station in 1945.

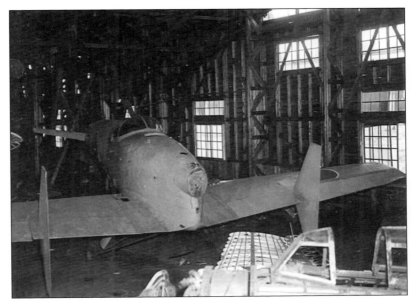

A rear view of the canard Yokosuka MXY6 research glider.

Below: two views of the Kyushu J7W1 Shinden 'Magnificent Lightning' canard fighter. *(USAAF)*

The Nakajima Kikka 'Orange Blossom' was Japan's first jet-powered aircraft. It was developed late in World War Two and the first prototype had only flown once before the end of the conflict. It was also called Kōkoku Nigō Heiki. The Japanese built 10, three of which were taken to the USA for evaluation.

aspect ratio, a tricycle landing gear and a large single-fin rudder. Power came from four 2,000 hp Nakajima NK9K-L 'Homare' 24 radial engines with Hitachi 92 turbosuperchargers driving four-bladed propellers. The engines were cooled by counter-rotating fans positioned just inside the engine cowlings. Defensive armament included power-operated nose, dorsal, ventral and tail turrets along with two free-swivelling machine guns at the beam positions.

What emerged was the Nakajima G8N Renzan 'Mountain Range'. The Japanese Navy designation was 'Type 18 land-based attack aircraft'. The Allied code name was 'Rita'. The initial prototype was completed in October 1944 and delivered to the Navy for testing in January 1945, a year after development had started. Three further examples were completed by June 1945, the third prototype being destroyed on the ground by US carrier aircraft.

Other than minor problems with the turbosuperchargers, the Renzan performed satisfactorily and the Navy hoped to have a total of 16 prototypes and 48 production-version G8N1s assembled by September 1945. But the worsening war situation and a critical shortage of light aluminium alloys led to the project's cancellation in June.

One proposed variant was the G8N2 Renzan-Kai Model 22, powered by four 2,200 hp Mitsubishi MK9A radial engines and modified to accept attachment of the air-launched Ohka Type 33 Special Attack Bomber.

Just prior to Japan's surrender in August 1945 consideration was also briefly given to producing an all-steel version of the aircraft, to be designated G8N3 Renzan-Kai Model 23, but the cessation of hostilities precluded any further development. After the war, one prototype was taken to the United States and scrapped after testing.

Chapter Twelve
Flights of Fantasy

No study of German advanced technology from the Second World War period would be complete without making mention of the myriad of appearances of so-called 'UFOs' made in UFOlogy, conspiracy theory, science fiction, and comic book stories - all of which make claims or tell of stories that link UFOs to Nazi Germany. These 'theories' describe supposedly successful attempts to develop advanced aircraft or spacecraft prior to and during World War Two, and further assert the post-war survival - and indeed development - of these craft in secret underground bases in Antarctica, South America, or the United States, along with their creators.

According to the information available on these, various potential codenames or sub-classifications of Nazi UFO craft such as *Rundflugzeug, Feuerball, Diskus, Haunebu, Hauneburg-Gerät, V7, Vril, Kugelblitz* - not to be confused with the self-propelled anti-aircraft gun of the same name - *Andromeda-Gerät, Flugkreisel, Kugelwaffe*, and *Reichsflugscheibe...* All have all been referenced in one form or another.

As early as 1950 accounts started to appear, possibly influenced by historical German development of specialised engines such as Viktor Schauberger's 'Repulsine' device. Elements of these claims have been widely incorporated into various works of fictional and purportedly non-fictional media, including video games and documentaries, often mixed with more substantiated information. German UFO literature often conforms largely to documented history on the following points:

- The Third Reich claimed territory in Antarctica, sent an expedition there in 1938, and planned others.
- The Third Reich conducted research into advanced propulsion technology, including rocketry, Viktor Schauberger's engine research, flying wing craft and the Arthur Sack A.S.6 experimental circular winged aircraft.
- Some UFO sightings during World War Two, particularly those known as 'foo fighters', were thought by the Allies to be prototype enemy aircraft designed to harass Allied aircraft through electromagnetic disruption; a technology similar to today's electromagnetic pulse weapons.

These speculations and reports were limited mainly to military personnel, and the earliest assertion of German flying saucers in the mass media appears to have been in an article which appeared in the Italian newspaper *Il Giornale d'Italia* in early 1950. Written by Professor Giuseppe Belluzzo, an Italian scientist and a former Italian Minister of National Economy under the Bennito Mussolini regime, it claimed that *'...types of flying discs were designed and studied in Germany and Italy as early as 1942'*. Belluzzo also expressed the opinion that *'some great power is launching discs to study them'*.

According to some sources Belluzzo was one of three specialists brought in by the SS to work on Rudolf Schriever's original *Flugkreisel* (Flying Gyro) disc-fan concept which was built and supposedly test flown sometime in 1943.

While the Schriever disc-fan concept worked, the prototype was unstable and like the smaller-scale BMW *Flügelrad* (Winged Wheel) series of disc-fans was prone to instability problems.

In 1953, when Avro Canada announced that it was developing the VZ-9-AV Avrocar, a circular jet aircraft with an estimated speed of 1,500 mph, German engineer Georg Klein claimed that such designs had been developed during the Third Reich. Klein identified two types of supposed German flying disks:

- A non-rotating disk developed at Breslau by V-2 rocket engineer Richard Miethe, which was captured by the Soviets, while Miethe fled to the US via France, and ended up working for Avro.
- A disk developed by Rudolf Schriever and Klaus Habermohl at Prague, which consisted of a ring of moving turbine blades around a fixed cockpit. Klein claimed that he had witnessed this craft's first manned flight on 14 February 1945, when it managed to climb to 12,400 metres in 3 minutes and attained a speed of 2,200 kilometres per hour in level flight.

British aeronautical engineer Roy Fedden is supposed to have remarked that the only craft that could approach the capabilities attributed to flying saucers were those being designed by the Germans towards the end of the war. UFOlogists often extract a quote from Fedden's papers. As we have already seen, he was the chief of the technical mission to Germany for the Ministry of Aircraft Production in which he stated: *'I have seen enough of their designs and production plans to realise that if they* (the Germans) *had managed to prolong the war some months longer, we would have been confronted with a set of entirely new and deadly developments in air warfare.'* Fedden is also quoted as saying that the Germans were working on a number of very unusual aeronautical projects, though he did not elaborate upon his statement. All of this is true, but as we have already seen, he was actually talking about discovered design, development or prototype projects, not super-advanced aircraft or spacecraft.

This whole area is one of claim and counter-claim, myth, legend, ego and impossible-to-trace accuracy of stories. Some accounts say one thing, others recall the same event but tell it from a totally different perspective, to the point that a second story doubts the existence of people referred to in the first! This is a fairly common occurance - and one that has been made doubly difficult by the process of 'sheep-dipping' Nazi scientists by the OSS during Operation Paperclip.

The coming of the world wide web has allowed everyone with a pet theory the space and the ability to stake their claim. Circular winged craft hidden in hangars built deep underground beneath Area 51 in Nevada... or perhaps top secret Nazi bases under the ice of Antarctica, where in 1938, the Nazis sent a large team of explorers - including scientists, military units and building crews on warships and submarines - to the Queen Maud Land region. While mapping the area, they discovered a vast network of underground warm-water rivers and caves. One of these caves extended down as far as 20-30 miles and contained a large geothermal lake. The cave was explored and construction teams were sent in to build a city-sized base, dubbed Base 211 or New Berlin, that hosted the SS, the Thule Society, serpent cults, various Nazi occultists, the Illuminati, and other shadowy groups.

At some point, the Germans either discovered abandoned alien technology or made contact with extraterrestrial explorers, described as

Greys or Reptilians. They learned or were taught how to replicate the alien technology, and used it to begin developing a number of super weapons including an advanced aircraft called an antigravity-disk, or flying saucer.

The story goes - and as far as I can tell, it started with a Bulgarian engineer named Vladimir Terziski who says he is the president of an 'American Academy of Dissident Sciences'. Terziski states that with their advanced rocket technology, the Nazis were able to launch a Moon mission in the early 1940s. They were able to create a base there because the Moon actually has a comfortable atmosphere that makes vegetation possible. No spacesuits are required! According to Terziski while many of these weapons were not ready for use in World War Two, the base and the ability to manufacture these weapons still exists and the Germans/aliens/some cult or secret society - depending on which conspiracy theorist you ask - will eventually launch a New World Order from it. Of course, the American moon landings were therefore staged to prevent the rest of the world from learning about the conspiracy - and it goes without saying that the Kennedy assassination and murder of Marilyn Monroe were all tied up it!

Die Glocke

Of all the so-called 'German UFOs' possibly the one most discussed and most well known due to public media exposure is *Die Glocke* - German for 'The Bell'. This is a purported top secret Nazi scientific technological device, secret weapon, or *Wunderwaffe*. Die Glocke has become a popular subject of speculation and has a following similar to science fiction fandom.

Discussion of *Die Glocke* seems to have originated in the works of Igor Witkowski. His 2000 Polish language book *Prawda o Wunderwaffe* (*The Truth About The Wonder Weapon*, reprinted in German as *Die Wahrheit über die Wunderwaffe*), refers to it as 'The Nazi-Bell.' Witkowski wrote that he first discovered the existence of *Die Glocke* by reading transcripts from an interrogation of former Nazi SS Officer Jakob Sporrenberg. According to Witkowski, he was shown the allegedly classified transcripts in August 1997 by an unnamed Polish intelligence contact who said he had access to Polish government documents regarding Nazi secret weapons.Witkowski maintains that he was only allowed to transcribe the documents and was not allowed to make any copies. Although no evidence of the veracity of Witkowski's statements has been produced, they reached a wider audience when they were retold by British author Nick Cook, who added his own views to Witkowski's statements in his 2001 book *The Hunt for Zero Point:Inside the Classified World of Antigravity Technology.*

Allegedly *Die Glocke* was an experiment carried out by Third Reich scientists working for the SS in a German facility known as Der Riese (The Giant) near the Wenceslaus mine and close to the Czech border, *Die Glocke* is described as being a device made out of a hard, heavy metal approximately 2.7 metres wide and 3.7 to 4.6 metres high, having a shape similar to that of a large bell. According to an interview of Witkowski by Cook, this device ostensibly contained two counter-rotating cylinders which would be '*...filled with a mercury-like substance, violet in colour'*. This metallic liquid was code-named Xerum 525 and was '*...stored in a tall thin thermos flask a metre high encased in lead'*. Additional substances said to be employed in the experiments, referred to as Leichtmetall (light metal), included thorium and beryllium peroxides.

Witkowski describes Die Glocke, when activated, as having an effect zone extending out 150 to 200 meters. Within the zone, crystals would form in animal tissue, blood would gel and separate while plants would rapidly decompose into a grease like substance. Witkowski also said that five of the seven original scientists working on the project died in the course of the tests.

Based upon certain external indications, Witkowski states that the ruins of a concrete framework - somewhat aesthetically nicknamed 'The Henge' - in the vicinity of the Wenceslas mine (50°37'43"N 16°29'40"E) may have once served as a test rig for an experiment in anti-gravity propulsion generated with *Die Glocke*. However, the derelict structure itself has also been interpreted to resemble the remains of a conventional industrial cooling tower - that aspect however does not often get told!

Witkowski's statements along with Cook's views prompted further conjecture about the device from various American authors, including Joseph P. Farrell, Jim Marrs, and Henry Stevens. In his 2007 book, *Hitler's Suppressed and Still-Secret Weapons, Science and Technology*, Stevens concludes that the violet mercury-like substance described by Witkowski could only be red mercury because normal mercury '...*has no fluid compounds according to conventional wisdom*'. Stevens goes on to present a story attributed to German scientist Otto Cerny as told to them by 13 year old Greg Rowe around 1961 which alleged that a concave mirror on top of the device provided the ability to see images from the past during its operation.

Witkowski stated that *Die Glocke* ended up in a Nazi-friendly South American country. Cook, on the other hand, speculates that it was moved to the United States as part of a deal made with SS General Hans Kammler. Farrell speculated that it was recovered as part of the so-called Kecksburg UFO incident that occurred in Kecksburg, Pennsylvania, USA on 9 December 1965. The same month, German engineer Rudolf Schriever gave an interview to German news magazine *Der Spiegel* in which he claimed that he had designed a craft powered by a circular plane of rotating turbine blades 15 metres in diameter. He said that the project had been developed by him and his team at BMW's Prague works until April 1945, when he fled Czechoslovakia. His designs for the disk and a model were stolen from his workshop in Bremerhaven-Lehe in 1948 and he was convinced that Czech agents had built his craft for an unspecified foreign power. In a separate interview with *Der Spiegel* in October 1952 he said that the plans were stolen from a farm he was hiding in near Regen on 14 May 1945. There are other discrepancies between the two interviews that add to the confusion.

The Kecksburg incident theory involving *Die Glocke* was dramatised in 2009 by The Discovery Channel and again in 2011 by The History Channel's *Ancient Aliens* series.

One only has to access the world wide web to see that stories relating to Nazi UFOs are virtually endless, as are the books and programmes that circulate endlessly on cable and satellite TV. Is there any truth to them? - I do not know.

All I can say is that in the many thousands of documents I have looked at over the years, I have yet to come across one device, one story or one shred of evidence that takes German developments beyond what has been described here.

Die Glocke may make good summertime beach reading, but as to factual evidence - it fails completely to ring my bell!

Bibliography and Further Reading

Aeroplane Monthly December 1994. Me.262 Jet Fighters

Aeroplane Monthly September 1993. Flying the Bf.109

Aeroplane Monthly March 2002. France's Dornier Do.335s Pt.1
Aeroplane Monthly April 2002. France's Dornier Do.335s Pt.2

Aeroplane Monthly June 2003. Me.262 Database

Aeroplane Monthly March 2004. Hendon's Me.262

Aeroplane Monthly May 2004. Arado Ar.234 Database

Aeroplane Monthly September 2004. Focke-Wulf Fw.190 Database

Aeroplane Monthly April 2006. The Salamander in France Pt 1
Aeroplane Monthly May 2006. The Salamander in France Pt 2

Air Pictorial and Air Reserve Gazette August 1954. Japanese Jet Aircraft at the end of the Second World War

American Raiders: The Race to Capture the Luftwaffe's Secrets by Wolfgang W. E. Samuel · University Press of Mississippi. ISBN 1578066492.

Something of a mystery picture - thought to have been taken at Forus, near Stavanger, Norway before being flown to the UK, this picture of an Arado Ar.234B is allegedly W.No. 140476.

Dark Side of the Moon: Wernher Von Braun, the Third Reich, and the Space Race by Wayne Biddle. W.W. Norton. ISBN 0393059103

Das Bell by Chris Berman. Leo Publishing, LLC. ISBN 1941157009

Dornier Do 335: An Illustrated History by Karl-Heinz Regnat · Schiffer Publishing, Limited. ISBN 0764318721

Dornier Do 335: The Luftwaffe's Fastest Piston-Engine Fighter by J. Richard Smith, Eddie J. Creek, Gerhard Roletschek. Classic ISBN 1903223679

Echo of the Reich by James Becker · Transworld Publishers Limited ISBN 0857500902

Fighting Hitler's Jets: The Extraordinary Story of the American Airmen Who Beat the Luftwaffe and Defeated Nazi Germany. by Robert F. Dorr · Zenith Press ISBN 0760343985

Focus: The Shaft of the Spear - Operation LUSTY: The US Army Air Forces' Exploitation of the Luftwaffe's Secret Aeronautical Technology, 1944-45. Daso, Dik Alan. 2002. Airpower Journal. 16, no. 1: 28.

Heinkel HE 162 Volksjager: Last Ditch Effort by the Luftwaffe by Peter Muller · Casemate Publishers ISBN 3952296813

Hitler's Flying Saucers: A Guide to German Flying Discs of the Second World War by Henry Stevens · Adventures Unlimited Press ISBN 1935487914

Hitler's Miracle Weapons: Secret Nuclear Weapons Of The Third Reich by Frederick Georg. Helion & Company. ISBN 978-1874622246

Japanese Secret Projects: Experimental Aircraft of the IJA and IJN, 1939-1945 by Edwin M. Dyer · Midland Counties Pubs. ISBN 1857803175

Jet Bombers: From the Messerschmitt Me 262 to the Stealth B-2 by Bill Gunston, Peter Gilchrist. Osprey Aerospace ISBN 1855322587

Last Hope of the Luftwaffe: Me 163, He 162, Me 262 by Maciej Góralczyk, Jacek Pasieczny, Simon Pasieczny, Arkadiusz Wróbel Kagero Oficyna Wydawnicza ISBN 8362878711

Luftwaffe Secret Projects: Ground Attack & Special Purpose Aircraft by Dieter Herwig, Heinz Rode · Midland Counties Pubs ISBN 1857801504

Luftwaffe X-Planes by Manfred Griehl Pen & Sword Books Limited ISBN 1848327897

A pair of Me.262s at Schleswig in June 1945.

Arado Ar.234B-2 at RAE Farnborough. The engines had been removed for inspection by Power Jets Ltd. *(Peter Green Collection)*

Man-Made UFOs: WWII's Secret Legacy by Renato Vesco, David Hatcher Childress · Adventures Unlimited Press ISBN 1931882770

Me 262 Bomber and Reconnaissance Units by Robert Forsyth, Eddie J. Creek, Jim Laurier · Osprey Publishing, Limited ISBN 1849087490

Me 262 Volume 2 by J. Richard Smith, Eddie J. Creek Classic Publications ISBN 0952686732

Me. 262 Stormbird Ascending by Bob Carruthers Pen & Sword Books Limited ISBN 1781592314

Messerschmitt Me 262: Variations, Proposed Versions and Project Designs Series Me 252 A-1a by David Myhra · Schiffer Publishing, Limited ISBN 0764319396

Most Secret War British Scientific Intelligence, 1939-1945. R V Jones. Hamish Hamilton, London 1978.

NASA, Nazis & JFK: The Torbitt Document & the Kennedy Assassination by William Torbitt, Kenn Thomas. Adventures Unlimited Press. ISBN 0932813399

Operation Lusty Harold Watson's "Whizzers" Went Hunting for German Jets-and Came Back with Several Jewels". Young, R. L. 2005. Air Force Magazine. 88: 62-67.
Operation Paperclip by Patrick Goddard. CreateSpace Independent Publishing. ISBN 099285900X

Operation Paperclip: The Secret Intelligence Program that Brought Nazi Scientists to America by Annie Jacobsen. Little, Brown & Co; New York 2014.

Reich of the Black Sun: Nazi Secret Weapons & the Cold War Allied Legend by Joseph P. Farrell. Adventures Unlimited Press. ISBN 1931882398

Tales from Langley: The CIA from Truman to Obama. by Peter Kross · Adventures Unlimited Press. ISBN 1939149169

The Black Sun: Montauk's Nazi-Tibetan Connection by Peter Moon · Sky Books · ISBN 0963188941

The Captive Luftwaffe – Kenneth S West. Putman & Co. London 1978,

The Hunt for Zero Point by Nick Cook · Random House Incorporated ISBN 0767914961

The Me 262 Stormbird Story by John Christopher History Press ISBN 0752453033

The Me 262 Stormbird: From the Pilots Who Flew, Fought, and Survived It by Colin D. Heaton, Anne-Marie Lewis MBI Publishing Company ISBN 0760342636

The Medusa File: Secret Crimes and Coverups of the U. S. Government. by Craig Roberts · CreateSpace Independent Publishing Platform ISBN 1495306690

The Race for Hitler's X-Planes: Britain's 1945 Mission to Capture Secret Luftwaffe Technology. by John Christopher. History Press ISBN 0752464574

The SS Brotherhood of the Bell: NASA's Nazis, JFK, and Majic-12 by Joseph P. Farrell · Adventures Unlimited Press ISBN 1931882614

The Vanishing Paperclips: America's Aerospace Secret. by Hans H. Amtmann · Monogram Aviation Publications. ISBN 0914144359

The World's First Turbo-Jet Fighter: Me 262 by Manfred Griehl. Schiffer Publishing, Limited ISBN 0887404103

The World's First Turbojet Fighter - Messerschmitt Me 262 by Heinrich Hecht · Schiffer Publishing, Limited ISBN 0887402348

UFO's Nazi Secret Weapons? by Mattern Friedrich by Mattern Friedrich, Christof Friedrich, Commander X · Inner Light - Global Communications · ISBN 1606111167

USAFM Friends Journal. Vol.14, No.1, Spring 1991 Wright Field and the German Me.262

Von Braun: Dreamer of Space, Engineer of War. by Michael J. Neufeld · Alfred A. Knopf · ISBN 0307262928

War Prizes - Phil Butler · Midland Counties Publications · Leicester 1994.

Watson's Whizzers: Operation Lusty and the Race for Nazi Aviation Technology · Wolfgang W E Samuel · Schiffer Military History 2010.

Wernher Von Braun: The Man who Sold the Moon by Dennis Piszkiewicz · Praeger · ISBN 0275962172

Wings of the Luftwaffe: Flying the Captured German Aircraft of World War II by Eric Brown · Hikoki Publications ISBN 1902109155

Index

A
Alberti, Dr Jack; US scientist: 220
Ali, Rashid, Iraqi politician: 29
Alles Kaput Ju 290:   179, 184
Amman, Dr BMW engine development: 125
Anspach, Lt Bob; USAAF pilot: 159, 160, 162, 166, 171, 174, 181, 182
Aranson, Flt Lt RAF: 174
Arnold, Ofw Heinz; German pilot: 162
Arnold, General Henry Harley 'Hap'; Head of USAAF: 6, 99, 104-107, 110, 181, 182
Atcherley, Group Captain David, RAF: 25

B
Baars, Ernst; former German chemist: 221
Bader, Group Captain Sir Douglas Robert Steuart; RAF pilot: 34
Badré, Colonel Paul; French test pilot: 224, 226, 227, 233
Baranov, A N; Russian scientist: 135
Barmin, V P Russian engineer: 137
Barnes, USS: 239
Barr, Flt Sgt RAF: 35
Bateman, Professor Harry; US scientist: 112
Baur, Karl German test pilot: 158, 159, 160, 162, 168, 173
Beeton, Flt Lt A B P RAF: 121, 126
Belluzzo, Professor Giuseppe, Italian scientist: 255, 256
Bennett, Flt Sgt: 43
Berger, Oblt Hans; German pilot: 233

Bertrand, Louis, French pilot: 233, 235
Beverly Ann Me.262:   162, 170, 182
Bialy, Squadron Leader; Polish Air Force: 37
Biot, Lt Cmdr M A; USN: 194
Blazig, Dr; German avionics specialist   140
Bock, Major; German officer: 69
Bogue, USS: 239, 241, 251
Boguslavsky, E. Yuri; Soviet technician: 137
Boitelet, Captain; French Air Force: 232
Bollay, William USN: 111
Both, Eberhard; former German chemist: 221
Bouvarre, Commandant; French Air Force: 232
Boyd, Colonel Al; USAAF: 185, 186
Brandt, Major Heinz; German officer: 63
Braur, Hauptman Heinz; German pilot: 154
Brewer, Dr A Keith; chemist: 220
Brill, Rudolf; former German chemist: 221
Broadhurst, Air Vice-Marshal Harry E; RAF: 90
Bromley, Major William; US Army: 128
Brown, Lt Cdr Eric Melrose 'Winkle'; RAF: 79, 90
Brown, Lt Roy; USAAF: 159, 160, 168, 171, 173, 179, 182
Bruckmann, Dr Bruno; BMW engine specialist: 123, 125, 126
Bulganin, N A; Soviet scientific specialist: 135

Bull, Gerald Vincent; Canadian scientist: 72
Burder, Sqn Ldr; RAF: 64
Busemann, Dr Adolph; German scientist: 221
Butler, Phil; author: 5
Buxton, Wing Commander Geoffrey Mungo; RAF: 77
Byrnes, James F; US politician: 218

C
Cameron, Maj. Gen. A M: British Army: 126
Campbell-Orde, Wing Commander I R; RAF: 22
Capper, Colonel John E; British Army 16
Caroli, Gerhard; Messerschmitt Aircraft: 158, 162
Cerny, Otto; German scientist: 258
Cheany, Flt Lt; RAF: 121
Chertok, Boris Evseevich; Soviet technican: 136
Christiansen, Lt Commander Hasager; Danish Naval officer: 51
Churchill, Winston Spencer; UK Prime Minister: 52, 63
Cody, Samuel F; American balloon pioneer: 16
Cohen, Kenneth; UK scientific advisor 52
Connie the Sharp Article Me.262: 161, 169, 182
Cook, Nick; UK Author: 257, 258
Cookie VII Me.262: 180, 182
Core, USS: 239, 240
Cripps, Sir Stafford; UK Minister of Aircraft Production: 121
Cross, Wing Commander V; RAF: 121

Cumberland Sound, USS: 240, 241
Czerniawski, Roman (Agent Brutus): 55

D
Dahlstrom, Captain Kenneth; USAAF: 171, 179, 182
De Havilland, Captain Geoffrey; UK aircraft designer: 16
Debus, Kurt H; rocket scientist: 216
Deilitz, Wilhelm Karl Otto; German aircraft technician: 177
DeLovely Me.262: 182
Dennis Me.262: 167, 183
Descaves, Captain French test engineer: 235, 237
Detts, Dr; German scientist: 113
Detweiler, Lt Colonel A; US Army: 176
Dispot, Capitaine; French aviation technician: 224, 226
Doblhoff, Dr German helicopter designer: 188, 189
Doenitz, Grand Admiral Karl: 214
Doris Me.262: 183
Dornberger, General Dr Walter: 63
Dornier, Claudiu Honoré Desiré; founder Dornier Aircraft: 224,
Dorsh, Xaver: 135
Dryden, Dr Hugh L: 106, 112
Dumke, Otto; German pilot: 177
Duncan, Professor Dr W J, Professor of Aeronautics, Hull University: 121
Duseman, Dr; German scientist: 114

E
Eckert, Dr; German scientist: 125
Eisenhower, General Dwight David; US Army: 51, 72, 127
Ellor, James E 'Jimmy'; UK engineer: 17

Arado Ar.234B-2 VK877 during it's time with the RAE. *(Peter Green Collection)*

Europa; German aircraft carrier: 8

F
Faber, Oberleutant Armin: 22-25, 39
Falck, Dr; German ship designer: 215
Falk, Wing Commander Roland John 'Roly'; RAF: 42, 43, 50, 79
Farrell, Joseph P; US author: 258
Fattler, Dr BMW engine scientist: 123
Fedden, Sir Alfred Hubert Roy; UK scientific engineer: 121-127, 256
Feudin 54th AD Sq Me.262: 182
Fleisher, Karl Otto; German scientist: 128
Flugzeugträger A; German aircraft carrier: 8
Flugzeugträger B; German aircraft carrier: 8
Forbes, Flying Officer R F; RAF: 31
Fourcade, Marie-Madeleine; French resistance worker: 52
Franz, Dr Ing. Anselm; Junkers Aircraft: 222
Freeman, Captain Richard S; USAAC: 102
Freiburger, M. Sgt Eugene E; US Army: 153, 154, 161, 162
Fuisting, Hans-Ehrenfried; German test pilot: 177

G
Gable, Captain Clark; Hollywood film star: 37
Gans, Richard; German scientist: 14
Ganzer, Vic; Boeing Aircraft: 207
Gentil, Captain; French flight test: 231
Georgii, Dr; German scientist: 113
Gerber, Eduard; former German engineer: 221
Gersenhaurer, Hans-Helmut; German test pilot: 175
Gibson, Professor A H; UK engineer: 17
Ginny H Me.262: 167, 170, 183
Girault, Jean; French test pilot: 226
Glantzberg, Colonel Frederick E. 'Fritz'; US Army: 110
Goddard, Colonel George; USAAF: 178
Goebbels, Paul Joseph; Reich Minister: 5, 7

Goering, Hermann; German politician: 195
Goubau, Georg; former German physicist: 221
Gough, Flying Officer G D M; RAF : 26, 28, 31, 41, 42, 50
Graf Zeppelin; German aircraft carrier: 8
Green, Major F M; UK engineer: 17
Griffith, A A; UK engineer: 17
Grimal, Colonel Georges, French BAT; 237
Grimes, Captain Clifton C: USN: 242
Groszkowski, Professor Janusz; Polish scientist: 65
Groves, Maj Gen Leslie; US Army: 108
Guenther, Richard; former German engineer: 221
Gurevich, Mikhail; Soviet aircraft designer: 90
Guttwein, Gunter; former German physicist: 221

H
Habermohl, Klaus; German flying disk designer: 256
Hamill, Major James P; US Army: 128, 218
Happy Hunter II Me.262: 161, 172, 182
Hards, Grp Capt Alan F; RAF: 42, 79, 86, 95
Hartkopf, Otto Junker Aircraft: 122
Hass, Georg; former German physicist: 221
Hawthorne, Sir William Rede; UK Scientist: 110, 111
Haynes, Lt William V; USAAF: 170, 183
Heinrich, Dr; German scientist: 125
Heisenberg, Werner; German physicist: 217
Helmschrott, Joseph; Messerschmitt Aircraft: 126
Henke, Captain Werner; German Navy: 217
Higgins, Staff Sgt; US Army: 153
Hillis, Captain Fred; USAAF: 168, 171, 173, 179, 182
Himmler, Reichsfuehrer SS Heinrich: 63, 92
Hirshauer, Commandant, French Air Force: 232
Hitler, Adolf; Reich Chancellor of

Germany: 8, 53, 61, 68, 69
Hoch, Dr; German avionics specialist: 140
Hoffman, Ludwig 'Willie'; German pilot: 158, 160, 162, 168, 170, 172
Holmes, Colonel Joel; US Army: 217
Holt, Lt Ken USAAF: 159, 160, 162, 166, 180, 181, 183
Holt, Lt. James K; USAAF: 167, 171, 179, 183
Hornung, Hans Messerschmitt Aircraft: 126
Horten, Major Walter German aircraft designer: 190, 195
Horten, Oberleutnant Reimar; German aircraft designer: 190, 195
Houtermans, Fritz; German scientist: 14
Howe, William E 'Bill'; US electronics engineer: 220
Hussein, Saddam; President of Iraq: 72
Huzel, Dieter; German engineer: 128

I
Inuzuka, Lt Commander; Japanese test pilot: 251
Iszkowski, Squadron Leader; Polish Air Force: 37

J
Jabo Bait Me.262: 168, 170, 178, 183
Jane 1 Me.262: 183
Jayne, Lt; USN: 194
Jeffreys, Sgt; RAF: 24
Joanne Me.262: 182
Jones, Professor Reginald Victor; UK scientist: 51-53, 62, 67, 72
Jones, R T; NASA scientist: 221
Jungert, Wilhelm; German rocket scientist: 217

K
Kammler, SS General Dr.-Ing. Hans (Heinz) Friedrich Karl Franz: 128, 136
Kedesdy, Horst; former German physicist: 221
Keeling, Sqn Ldr; RAF: 42
Kennedy, Joseph P Jr, USN 72
Kennedy, President John F: 72
Kenworthy, John; UK Engineer: 17
Kerimov, Kerim; Soviet engineer: 138
Kersting, Herman; German test pilot: 158
Khrulev, A V; Soviet scientist: 135
Kinder, Flying Officer; RAF: 31
King George VI: 37
King Hathaway, SS: 197
King, J C; RAE: 121
Kinzy, R E; aeronautical engineer: 220
Klaus, Samuel; US politician: 218, 219
Klein, Georg; German engineer: 256
Klien, Dr Arthur L; US scientist: 113
Korolev, Sergei; Soviet scientist: 133
Kracht, Felix; German aircraft designer: 80
Kronfeld, Robert; UK pilot: 90
Kuznetsov. General Viktor; Soviet aviation specialist: 138

L
Lady Jess IV Me.262: 171, 182
Lafferenz, Otto; Director, Deutsche Arbeitsfront: 12
Lambert, René, French flight engineer: 229, 230
Le Martelot, Captain; French test engineer: 235, 237
Lee, Flt Sgt; RAF: 41
Lehovec, Kurt; former German physicist: 221
Lewendon, Flt Lt E R; RAF: 28, 31, 38, 39, 41
Ley, Willy; German scientist: 62
Lindbergh, Charles A; US aviation pioneer: 109, 160
Lindemann, Professor Frederick Arthur ; scientific advisor to UK Government: 52
Lombaert, Lt Raphael; French pilot: 235, 237
Longuet, Lt Col, Director CEAM: 237
Lutz, Otto; German engine designer: 125

M
Mackle, Unteroffizier Hans; German pilot: 46
Magnus, Kurt; German engineer: 140
Malenkov, G M; Soviet technician: 135
Marge Me.262: 182
Marrs, Jim; US author: 258
Martin, George; Boeing Aircraft: 207
Martindale, Sqn Ldr A F; RAF: 42

He.162A VH513 displays over RAE Farnborough. *(Peter Green Collection)*

Matsukasa, Admiral, Japanese Navy: 243
Maurin, Commandant Philippe; French Air Force: 237
McAuley, Major Walter J: 184, 185, 186, 187
McCarthy, Sqn Ldr Joe; RAF 79
McDonald, Brigadier General George C; US Army: 172
McIntosh, Captain Fred B; USAAF: 150
Merifield, Sqn Ldr John; RAF: 55
Messerschmitt, Professor Willi; German aircraft designer: 167
Mikoyan, Artem; Soviet aircraft designer: 90
Millikan, Dr Clarke Blanchard; US scientist: 112, 113, 118-120
Mirles, Captain; French Flight test: 231
Missouri, USS: 239
Mitchell, Reginald Joseph; UK Aircraft designer: 34
Mockle, Obgefr. Hans;Luftwaffe: 46
Mollison, Flight Captain James; Air Transport Auxilliary: 38
Morch, Commodore Paul; Chief Danish Naval Interlligence Service: 51
Mozhorin, Yuri; Soviet scientist: 138
Muller, Lt Fritz; German pilot: 162
Multhopp, Dr Hans; German aircraft designer: 88, 89, 90, 221
Mussolini, Benito Amilcare Andrea; Italian Prime Minister: 28, 255

N

Nebel, Rudolf, German rocket scientist: 62
Neubert, Erich W; German rocket scientist: 217
Newport, Bert, UK engineer: 121
Nicolle, Abel; French pilot: 233, 235
Niki, Captain; Japanese aircraft designer: 246
Northrop, Jack; US aircraft designer: 190
Nosovsky, N E; Soviet scientist: 135

O

O'Gorman, Mervyn; UK engineer: 16
Oberth, Hermann; German physicist: 14, 62,
Olbricht, General Friedrich; German officer: 63
Ole Fruit Cake Me.262: 173, 183
Olze, Obgefr. Heinz; Luftwaffe aircrew: 46
Osenberg, Werner; German scientist: 216
Osterwald, Heinrich; test pilot: 232

P

Patterson, Lt. Frank Stuart; USAAC: 97
Patton, General George S; US Army: 14
Pauline Me.262: 182
Peek, Flt Sgt E P H; RAF: 62

Peter Strasser; German aircraft carrier: 8
Petit, Captain; French Air Force: 232
Petrov, General Nikolai; Soviet scientist: 136
Pick II Me.262: 180, 182
Pobedonostsev, Yuri. A; Soviet scientist: 137, 138
Poppel, Theordor A; German rocket scientist: 217
Popson, Captain Raymond A; US test pilot: 190
Portal, Marshal of the Royal Air Force Charles Frederick Algernon: 52
Prandtl, Professor Ludwig; German scientist: 113, 115, 122, 221
Pujol, Juan (Agent Garbo): 55, 56

Q
Queen Elizabeth: 37
Quill, Jeffrey; US test pilot: 34

R
Raeder, Grand Admiral Erich: 8
Randrup, Sqn Ldr; RAF: 42
Reaper, HMS: 101, 167, 172, 174, 176-182, 183
Receveau, Roger; French test pilot: 228, 229, 230
Rees, Eberhard; German rocket scientist: 217
Reid, Flt Lt; RAF: 121
Reiman, Ernst-Willi; German pilot: 177
Reitsch, Hanna; German test pilot: 59
Remizov, Colonel Yuri T; Soviet technician: 139
Rickhey, Georg; rocket scientist: 221
Ridel, Walther; German scientist: 128
Rowe, Greg: 258
Rozanoff, Colonel Kostia, Director CEAM: 233, 237
Rudolph, Arthur; rocket scientist: 216, 221
Ryazansky, Mikhail S; Soviet technician: 137, 138

S
Saburov, M S; Soviet scientist: 139
Sachse, Dr German engineer: 125
Sandys, Duncan; UK politician: 72
Sauckel, Ernst Friedrich Christoph 'Fritz': 132
Schaaf, Dr; German engineer: 123, 124
Schade, Capt Henry A: 108, 109
Schairer, George; Boeing Aircraft: 207
Schauberger, Viktor, engine designer: 255
Schiebold, Ernst, German scientist: 14
Schleeh, Russ; US test pilot: 183, 184, 185
Schlicke, Dr Heinz, German electronic expert: 214, 215
Schlienger, Captain; French pilot: 238
Schmellenmeier, Heinz, German scientist: 14
Schmidt, Dr; German scientist: 118, 125
Schmidt, Wulf (Agent Tate): 55
Schriever, Rudolf; flying designer: 256
Schuirmann, Rear Admiral Roscoe E: 107
Schulze, August; German rocket scientist: 217
Schwesinger, Gerhard; former German optician: 221
Schwidetzky, Walter; German rocket scientist: 217
Screamin Meamie Me.262: 170, 182
Seashore, Lt Colonel 'Bud'; US Army: 178
Serov, Colonel General; Soviet NKVD: 140
Shakhurin, A I; Head of Soviet aviation industry: 138
Sheldon, Colonel Huntington D 'Ting'; US Army: 170
Sieber, Lothar; German test pilot: 77
SNAFU Me.262: 180, 183
Soestmeyer, Christoph; BMW designer: 126
Sokolov, General Andrei Illarionovich, Soviet technician: 137
Sommerfield, Major, German technical advisor: 52
Spaatz, General Carl; USAAF: 51, 141, 170, 171, 172
Speer, Albert; German politician: 61, 67, 68
Spiegel, Dr; BMW engine designer 125
Sporrenberg, Jakob; SS Officer: 257
Stalin, Joseph; Premier of the Soviet Union: 6, 135
Staples, Flying Officer; RAF: 41
Staver, Major Robert B; US Army: 217

Staver, Major Robert; UK rocket technician: 128
Steenbeck, Max; German scientist: 14
Stempfie, A; Messerschmitt Aircraft: 202
Stern, W J; Allied Control Commission: 121, 126
Stevens, Henry; US author: 258
Stever, H Guyford; US scientist: 112
Stiemer, Oblt Gerhart; German pilot: 233
Stoffergen, Dr; BMW Aircraft Engines: 123
Strobell, Lt Robert C; USAAF pilot: 150, 152, 153, 159, 160, 162, 170, 171, 181, 182
Strughold, Hubertus; German physician: 216, 221
Stüper, Dr German scientific technician: 125
Sutcliffe, W; RAE: 19

T
Tanegashima, Dr Tokutsau, Japanese Navy: 243, 244
Tank, Kurt; German aircraft designer: 89
Terziski, Vladimir; Bulgarian engineer: 257
Thiel, Dr Walter; German rocket designer: 62
Thierry, Coammandant Emile; French Air Force: 237
Tirpitz; German battleship: 8
Toftoy, Colonel Holger Nelson ; US Army: 127, 217
Trejtnar, Sergeant František; Czech pilot: 23
Triumph, HMS: 90
Trubachev. Pavel; Soviet rocket technician: 138
Truman, President Harry: 215, 216, 218, 219
Tsetsior, Zinovy; Soviet rocket technician: 138
Turner, Flt Lt; RAF: 42
Tyulin, Lt. Colonel Georgy; Soviet rocket technician: 137, 138

V
Vakhitov, Lt General F I; Soviet rocket technician: 135
Valier, Max; German rocket scientist: 62
Vera Me.262: 161, 168, 182
Vogt, Dr Richard; aircraft designer: 221
Voight, Dr Waldemar; Messerschmitt designer: 126, 198, 200, 221
Voltaggio, Frank Jr; US test pilot: 183, 184
Von Braun, Wernher Magnus Maximilian; German rocket scientist: 62, 63, 128, 129, 214, 216, 217, 220
von Kármán, Dr Theodore; US scientist: 105, 106, 110-113, 221
Voznesensky, N A, Soviet rocket technician: 135

W
Wachtel, Colonel Max; German officer: 52, 58
Wagner, Professor Herbert A; Henschel Aircraft: 163
Ward, Captain; USAAF: 161
Watson, Colonel Harold E; USAAF pilot: 101, 145, 149, 152-154, 158-161, 166-170, 172-174, 179, 181-183
Watson, Lt Cdr C H; USNR: 107
Weickmann, Helmut; former German geophyiscist: 221
Weightman, Sqn Ldr; RAF: 42
Wellwood, Flt Lt; RAF 42
West, Kenneth S; author: 5
West, Major Carroll R; USAAF: 242
Wev, Captain Bosquet N; Head of JIOA: 218, 219
What is It? Me.262: 183
Whittle, Frank; jet engine designer: 111
Wideroe, Dr Rolf; German scientist: 14
Wigand, Friedrich; former German scientist: 221
Wilkinson, Sqn Ldr G S; RAF: 17, 64
Will, Frederick; : 177
Williams, Lt Col R L; US Army: 217
Willie Me.262: 161, 168, 172, 182
Wilma Jeanne Me.262: 160, 161, 182
Wilson, Group Captain Hugh John 'Willie'; RAF: 21, 25, 26
Wilson, Squadron Leader L D; RAF: 22
Winjah, USS: 172
Winkler, Johannes; German rocket experimenter: 62
Winter, Dr Martin; German aircraft designer: 88

Winterberg, Friedwardt; former German scientist: 221
Witkowski, Igor; Polish author: 257, 258
Wocke, Dr Hans German aircraft designer: 74
Wollenwebner, Oblt Wolfgang; German pilot: 233
Woodruff, Dr Louis; US scientist: 128
Woods, Robert J; Bell Aircraft: 200
Woolams, Jack; US test pilot: 93, 180, 181
Wurster, Dr Hermann; German test pilot: 158

Z

Zelewsky, Heinz; German pilot: 177
Zernov, P M; Soviet rocket technician: 135
Zhukov, Marshall Georgy Konstantinovich: 139
Ziegler, Hans; former German engineer: 221
Ziller, Lt Erwin; German test pilot: 193
Zwicky, Dr; V-2 rocket technician: 116